CREATIVE

MACHINE

DESIGN

Creative Machine Design

DESIGN INNOVATION
AND THE RIGHT SOLUTIONS

by Ben-Zion Sandler, Ph.D.
BEN-GURION UNIVERSITY OF THE NEGEV, BEER SHEEVA, ISRAEL

The Solomon Press NEW YORK

The Solomon Press
Publishers Creative Services Inc.
89-31 161 Street; Suite 611
Jamaica, New York 11432
United States of America

The author and publisher wish to thank the copyright owners who have given their permission to use copyrighted material. Any ommissions or errors in giving proper credit are unintentional and will lbe corrected at the first opportunity after the error or omission has been brought to the attention of the author or publisher.

Library of Congress Cataloging in Publication Data

Sandler, B. Z., 1932-
Creative machine design.

Bibliography: p.
Includes index.
1. Machinery—Design. I. Title.
TJ233.S25 1985 621.8'15 85-11964

ISBN: 0-934623-00-7

TABLE OF CONTENTS

LIST OF TABLES

LIST OF FIGURES

ACKNOWLEDGMENTS

I express my deep gratitude and appreciation to my colleagues at the Mechanical Engineering Department of the Ben-Gurion University of the Negev for their spiritual support and cooperation in accumulating the material for this book and assistance in writing it.

I am especially grateful to those who transformed my poor Russian English into something readable, particularly Inez Muerinik. I wish to thank Evelyn Tucker of The Solomon Press for her excellent copyediting, as well as the many who contributed to the editing, who typed, and to those who offered their advice in developing the manuscript. I thank Dr. Vladimir Maxson who made many valuable suggestions. My appreciation to Morion Milner for her valuable assistance. And, of course, many, many thanks to my wonderful family and my wife whose warmth and humor help me in all aspects of my life.

PREFACE

In my work as a professional engineer, designer and teacher, I have had many inquiries about specific machine design problems and general inquiries about the creative process as it applies to this area of engineering. These questions have come from fellow engineers, faculty colleagues and from students.

This book was conceived as a result of my own frustration at not being able to refer people to a book that would have an answer to these vital questions. In developing the content of the present work I have highlighted the creative-imaginative aspects and related this to the creative-practical side of engineering design. I have tried to keep in the forefront the end-result of the design process, namely the creation of a new, useful and saleable product.

Part I deals with the methodology for seeking technical solutions to design problems, the development of ideas and concepts, and the organization and execution of creative thinking. Problems are described as assignable to two planes: "what" to design and "how" to design it. The former are of interest to salespeople, marketing executives, and the like, and the latter to engineers and designers. It encompasses Chapters 1–4.

In Part II the reader is led directly into the stages of design after the "what" problem has been solved. Here the pathway from engineering principles to design is charted. This section consists of Chapters 5–9.

5. The design of the processing layout of the machine or device, in which the principles of action and the operation sequences are defined.

6. The design of the kinematic layout of the machine or device, which defines the way by which the elements are caused to move in the desired order.

7. A list of rules, conventions, and methods relating to the design of machine parts. A wide range of situations is illustrated for which the designer must know how to provide satisfactory reliability of the machine part. No classic design course treats these aspects of design. Nevertheless knowledge of the rules is very important in the machine design process, and this information can be useful not only to a beginner in the field, but also to the experienced engineer.

8. The use of computers in technical creativity; a discussion of the nature of creativity and how it can be computerized—if at all.

9. Some marketing problems.

Academic reasoning, which can cause difficulties for inexperienced readers, is kept to a minimum. Highly detailed classification is more suitable for the specialist than for the beginner, and thus example is preferred to discussion. The examples are simple enough to illustrate clearly the development of the engineering thought process. They do not require long explanations, in contrast to those presented in some books in the field, which suffer from the drawback of being too specific and professional. The cases described here should be familiar to most students.

The text includes a number of exercises and questions on different types of creative thinking, most of which have been tried out in class, in various projects, or in the laboratory.

We believe that in this book we have achieved an important objective—to introduce the beginner in engineering to the main approaches to the different stages of design. We have tried to accomplish this by classifying the methods involved in the creative search for technical solutions.

CREATIVE

MACHINE

DESIGN

Part I

INTRODUCTION

The past 150–200 years have been years of incredible progress in technology. And the main criterion on which such an evaluation is based is one that is easily measurable—it is productivity or efficiency in industry and manufacturing, or more simply, the number of units produced per unit of time. Thousands of examples can be cited to illustrate this tremendous progress. Here are just a few.

Today metal is cut at a velocity of 15 kg of chips per minute at a relative speed between the cutter's edge and the blank of 500 meters per minute. In 1825 the same parameters were about 0.15 kg of chips per minute at a speed of 5 meters per minute. Writing is another example. A modern typewriter can reach a speed of about 300 typed symbols per minute, and produce some ten copies simultaneously. In addition, with modern copying equipment an original can be reproduced many times per minute, exactly and ready for use. Compare this achievement with the production of books and manuscripts by hand 500 years ago. The classical example of progress, however, is weaving. The output of a modern loom is about 300–500 woofs per minute versus the 10–20 woofs per minute produced a hundrd years ago.

The most impressive technological achievements of the past two centuries have been in the realm of transportation—on the surface of the earth, on the sea, in the air, and, of course, in space. Before the advent of the steam engine, practically the only means of maintaining communication between geographical points on land was by horse. A team of horses can attain a speed of almost

30 kilometers per hour in a relatively short time, with the average velocity 10–15 kilometers per hour. A journey by sailing ship from Europe to America took six to ten weeks and was completely dependent on the weather and the winds, but ships were the only means of transportation available to those early travelers across the ocean. The same limits existed for communications: cablegrams, telegrams, air mail, and radio are all "youngsters" in terms of history. During the past 200 years, the speed of delivering information has increased by at least 50–100 million times. Compare the time required for a telephone call from the United States to Europe with that to deliver a letter via a sailing vessel, or even a steamship.

We could continue giving examples indefinitely, but even the few described here should lead to the conclusion that while technology has leapt forward, the productivity of thinking, especially creative thinking, has barely increased. The brain of a modern man or woman does not evaluate situations much faster than did the human brain 100, 200, or even 1000 years ago. It is true that people today know much more and are equipped with many kinds of technical devices, but the situations with which people now have to deal are much more complicated and cumbersome than those of the past. As a result, we often make decisions that are ridiculous. To quote only one example: To modify an item such as a modern machine tool to suit the production of a new product may take the production staff of a factory six months, and to design a new machine may take a year or two. This time is somewhat longer than the ever-changing market for tool machines will allow. In those two years, new concepts can appear and the new machine may be hopelessly obsolete at the very moment of its birth. Consequently it may be better not to initiate the process of implementing the new product in the factory. The design time is critical here. One may offer the objection that today we have sophisticated computing equipment that enables us to obtain information about hundreds of variations within a few hours, including all the auxiliary time-intensive operations connected with the system. Gear-speed reducers, for instance, can be completely calculated in minutes, including the time required for walking to the computer terminal room, punching the cards, and returning with the sheets bearing the calculation results. The process can be further speeded

up with the help of a auxiliary devices, such as aids for pencil sharpening, erasing incorrect lines on drawings, and copying. Modern desks equipped with rulers that are firm, straight, and perpendicular are available, and these considerably ease the work of the drafter. We also have computer-operated design systems in which the computer actually prints out the drawing of the object being designed. These systems have the ability to present tens, or even hundreds, of visual variations of the design pattern in several spatial positions in a few moments. This is of great importance for automobile and aircraft design and for other fields involving machine design and building.

However, all these "miraculous" mechanisms can be utilized only after a certain stage of design or thinking. Before these heavy weapons can be brought out, it is necessary to know what is going to be designed and what principles will be used for this purpose. We do not, as yet, have any automatic system that can invent a device, create a new product, or propose a new principle. Even today this part of the creative process must be performed by human beings. What can help us at this stage is possibly an effective "thinking organization" and the knowledge of the basic "laws" that control this process.

Although many mechanisms, devices, and methods look familiar, are well known, and have been used for a long time, someone may suddenly propose an original and unexpected solution to a particular problem. Ask how this splendid way of solving the problem was arrived at and very seldom will its inventor have a definite answer. Mostly the answer will take the form: "I thought, it took a long time, and *suddenly*!" What is this "suddenly"? It is the problem with which the first part of this work will deal.

It appears that the process of developing a new product occurs on two planes. The better known plane is related to typical engineering calculations, and its basics are taught at educational institutions. As a result of scientific research, this plane develops rapidly. Each year reports and scientific conferences add new information to this computative domain. However, the other plane of engineering activity—the process of creative seeking of new ideas, solutions, and approaches—must precede the calculations. Before one can begin to calculate, one must be able to create a model or to have a picture of the object under design. This latter

domain will constitute the subject of this work. The engineer or inventor has to overcome a number of obstacles during the design process—"what" to develop and "how" to go about it.

The solution to the problem may be found with the help of Figure I-1. Here we see a ream of sheets. Each sheet represents a coordinate plane in which we call one axis "what" and the other "how." Each sheet represents a separate design stage. Let us illustrate the use of this figure with an example—the creation of an electromobile (an automobile driven by electric power).

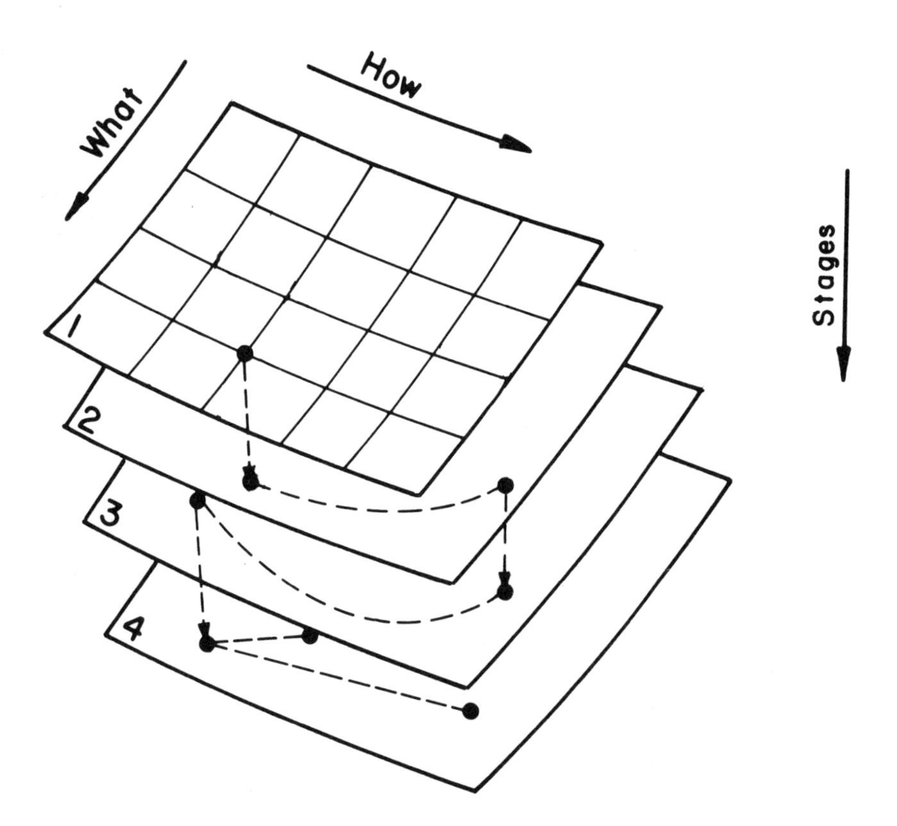

Figure I-1 The "what" and "how" model of technical creativity. Each sheet represents a coordinate plane in which one axis is denoted "what" and the other "how." Each sheet also represents a specific design stage. The dotted line is the trajectory of the design process.

For the first stage or sheet, one has to define *what* is to be designed and *how* to operate it:

What—a self-driven vehicle.
How—driven by a horse, a steam engine, an internal-combustion engine, an electromotor, and so on.

Once the choice to use an electromotor has been made, one passes on to the second sheet. Here:

What—an electromobile.
How—source of electricity can be chemical batteries, accumulators, solar cells, an autonomous generator, ultrahigh-frequency supply, and so on.

Once the decision to use an accumulator has been made, one passes on to the third sheet:

What—an electromobile with an accumulator-type energy source.
How—accumulator can be alkaline, acidic, or something unknown, charged from a stationary energy source located at home, from distribution points cited on the streets, or from charging stations.

After that choice has been made, one proceeds to the fourth sheet:

What—an electromobile with an alkaline accumulator that may be charged at home.
How—for the building of the car, how to design the driving mechanisms and the energy sources, and how to ensure the necessary parameters such as load, speed, and distance.

As one passes on to the following sheets, the solutions and details of the design become more specific and detailed. Each step is realized as a result of long and deep thinking.

The creativity factor appears in both the "what" and "how" coordinates on each sheet. Each sheet places an enormous stress

on the intellectual power of the designer, who attempts to make the best decisions at each stage, and it may take months, or even years, to go through the ream. This process has not significantly changed during the past 50, 100, or 200 years. It is no small wonder that attention is being paid to ways of speeding up the process of creative thinking.

The "what" problem encompasses a wide spectrum of questions, including the marketing, economic, and manufacturing possibilities at our disposal.

What should we develop?
What should we produce?
What will be in demand?
What will bring the best profit?
What will attract the customers' interest?

The most unpredictable solutions are possible here, but the problem of how to stumble on them remains. This is the problem of generating new ideas; it is the problem of speeding up the generating.

To solve the "how" problem, a wide range of knowledge in physics, chemistry, and technology is required. Much scientific erudition is the result of the striving to find an optimal solution to a previously formulated task. Here lies the source of the engineering calculations that are not the current object of our interest. We shall consider only the first, creative part of the "how" problem, although there is obviously always an interaction between "what" and "how" and the chain develops from "what" through "how" to calculation.

Chapter 1

PRINCIPLES OF CREATIVITY IN ENGINEERING

The examination of a familiar example should help to understand the role of creative thinking in the principles of design. The process of weapons' development is such an example. Unfortunately weaponry is a subject of common knowledge: each of us is familiar with its main concepts from childhood. It is, therefore, not difficult to define the requirements a weapon must fulfill:

1. To hit the target from as far away as possible.
2. To hit the target as strongly as possible.
3. To hit the target as accurately as possible.
4. To hit as many targets as possible at one time.
5. To conform to any economic restraints that may be placed on the weapon's production.

Having neither patent offices nor educational establishments at their disposal, primitive people used stones, pieces of wood, or sticks to throw at enemies or at animals they wanted to kill for food.

THE PRINCIPLES OF CREATIVE DESIGN

The First Principle

We should now be ready to formulate the first principle of creative thinking in design.

The designer copies or duplicates the events, phenomena, or processes taking place in nature, in neighboring fields of human activity, or even in far-removed fields of activity. Let us look first at the "copying" nature of machines. Here it would certainly be wrong to assume that borrowing from nature was typical only of ancient times. For example, the design of one of the most modern navigation devices is based on a navigation principle used in nature by the beetle. This insect finds its way by means of a "tuning fork," which excites the ganglion responsible for the direction of flight when the plane of the "tuning fork's" oscillations changes. The steam engine is another example of this principle: Many inventors tried to design their engines in a horse-like form. The mere fact that the engine was created to pull a load instead of to push it confirms this idea. The way in which the steam engine mimics nature was beautifully described by H.G. Wells (1866–1946), who said that the shadow of a horse runs in front of every steam engine.

If we think about "borrowing" from adjacent fields of human activity, we see that many modern machines and tools have hand-operated predecessors. The inventors of these machines tried to duplicate the manual process, for example, in the sewing machine and the loom. Although the needle and thread were tools, the sewing process was manual. Similarly, for thousands of years weaving was a completely manual process.

Over the years hunters learned to shape their stones to get sharp edges or to sharpen their stick to inflict greater damage on their enemies or prey. When metals were "discovered," these hunting tools were transformed into knives and spears. The steps described conform to the second requirement—to increase the effectiveness of the thrust. As the skill and experience of the ancient hunters and warriors grew, they came to understand that a straight stick hits more precisely and flies further and that a sharpened stone cuts better. They also found that by shaping a stick in a specific way, it could be made to return to the hunter's feet if it missed the prey—an important property, since it saves energy and decreases the danger to the owner of the weapon. We are still familiar with this kind of hunting tool in the boomerang.

Today we often apply the expertise gained in one field to the development of a device in another. For example, the idea of using a reversed wing in the design of sports cars was "borrowed" from aviation. (Figure 1-1). This modification provides the car with an

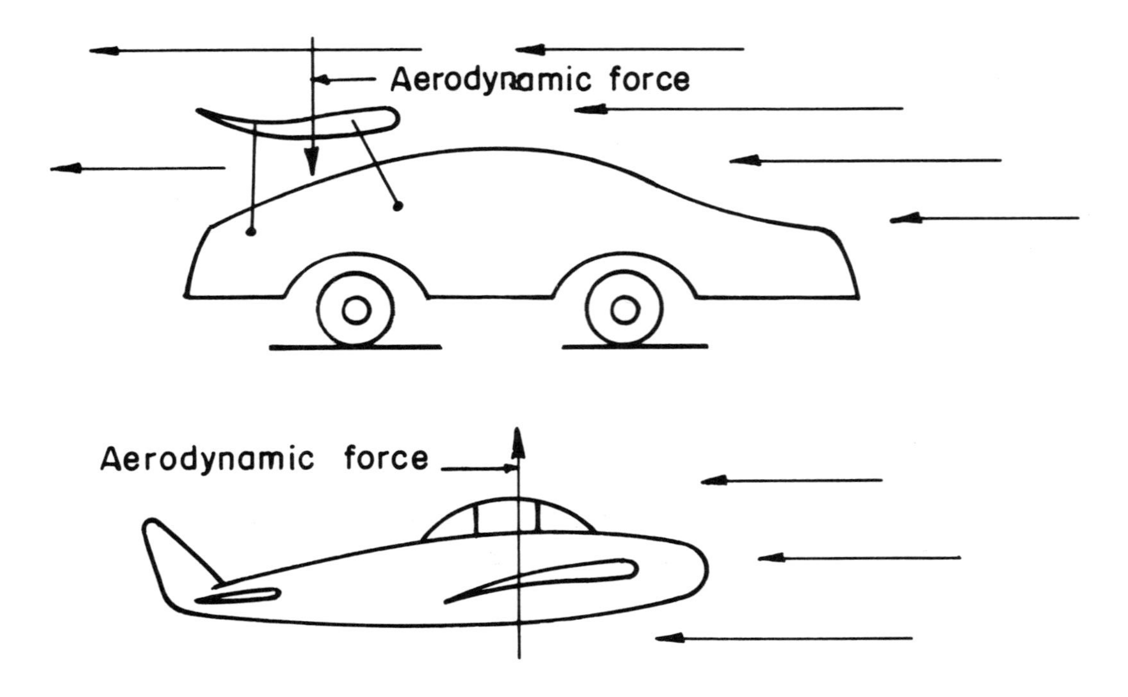

Figure 1-1 The reversed wing "borrowed" from aircraft design provides an automobile with an additional force that presses the wheels onto the road at high speeds.

additional force that presses the wheels onto the road at high speeds. The aerodynamics of the wing creates a force which, in this case, is equal to the carrying capacity of the wing and is directed toward the ground.

The Second Principle

The second principle of creativity can also be derived from the examples given, and can be stated as follows:

II. The existing features of a tool, device, or product are improved by streamlining and emphasizing the most important features. Sharp edges, better materials, and smoother surfaces of the stone, stick, and knife made them more effective and more dangerous tools. The better the streamlining of the throwing tool, the deeper it penetrates in spite of the fact that the initial value of mV (m mass, V velocity) does not change. In addition, the better the material, the longer the tool will last. As the use of new materials becomes widespread, new shapes sometimes become available, and the dimensions of many devices also change as technology becomes more sophisticated. An increase in the capacity of tools, weapons, or devices is often achieved by increasing their dimensions. For example, people began to build bigger bows and to use stronger materials, including steel, which led to the production of arbalests and ballistas. Similarly, after the introduction of gunpowder into military technology, there was a tendency to increase the dimensions of guns. Toward the end of World War I, for example, the Germans built their "Big Bertha" to shell Paris. (The results were disappointing from a military point of view.) Conversely, as technology develops, dimensions sometimes tend to decrease. The history of timepieces is an excellent illustration of this process, ranging from traditional large, tower clocks to today's tiny wristwatches.

The application of the second principle thus leads to the development of new products and the creation of new tools. It also involves the possibility of combining different tools to produce new ones. For instance, the spear was created from a stone or a metal tag fastened to the edge of a stick, and the ax by a differently shaped stone or metal body fastened to a thicker stick.

A classic example of new devices created by combining two existing technologies is the family of "geared linkages." These devices consist of a combination of a linkage mechanism, for instance, a four-bar mechanism, and a gear transmission. These new mechanisms have very interesting and important kinematic properties, as shown in Figure 1-2. In this figure the wheel 1 is the driving link. It rotates around its center 0_1. As a result of the action of links 3–4 and of a block of geared wheels 5, the wheel 2 is driven around its center 0_2. The ratio of this mechanism and the angle ϕ_2 as a function of the angle ϕ_1 are shown in the accompanying graph. We can see that there is an interval $\Delta\phi$ during every revolution of wheel 1 when wheel 2 is practically motionless. This property is very useful in some cases.

The Third Principle

We can now formulate the third principle of creativity as follows:

III. New tools and products can be created as the result of a combination of known elements. We are perhaps now approaching the most important point. The excellently shaped ancient tools were the outcome of the skill and experience of generations of craftsmen, their successes and failures. But no matter what they did, a sword remained a sword and a spear was still a spear. A new concept was necessary to push the development of the weapon further, and this modification was achieved by separating the weapon from the hands of the hunter or the warrior. The bow and arrow and the sling are the most important examples of this family of weapons (Figure 1-3).

The Fourth Principle

Let us now formulate the fourth principle.

IV. At a certain stage in the development of a new approach to the solution of a problem, it becomes necessary to introduce further advances in the product under consideration. What is the big difference between the bow and its predecessors? A hunter

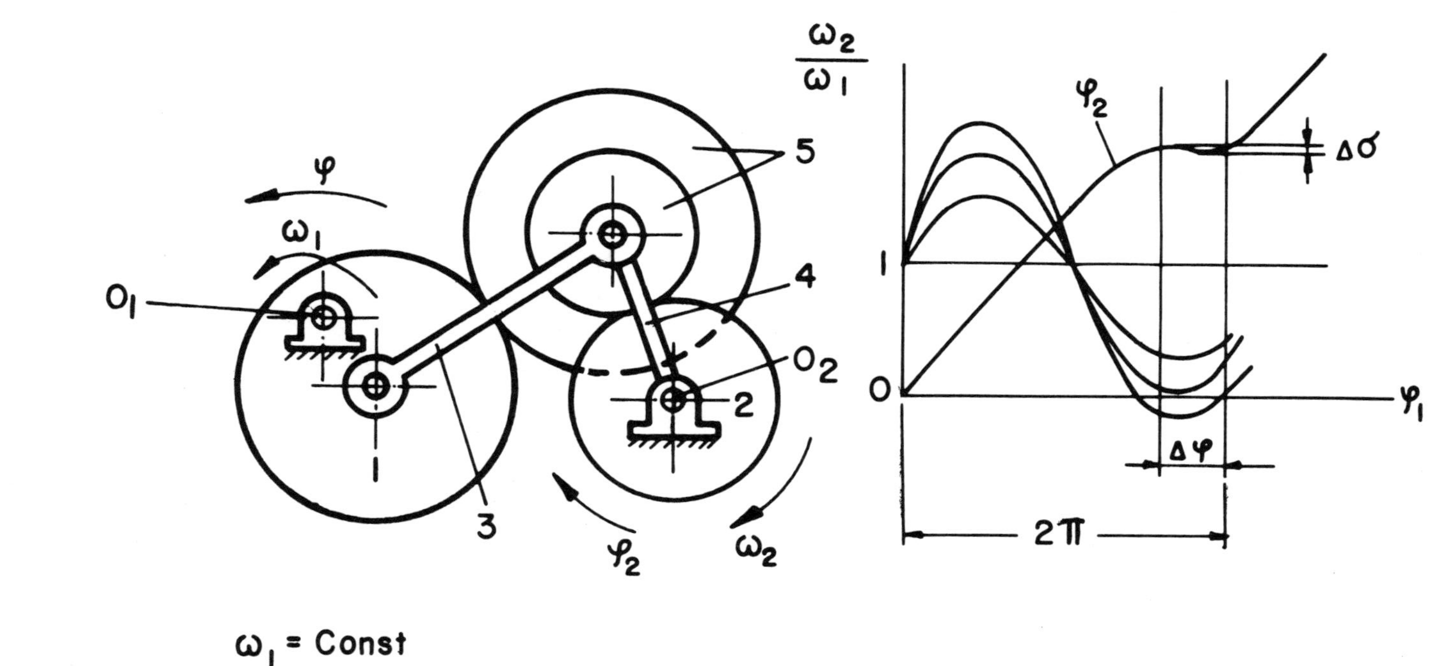

Figure 1-2 A four-bar linkage and a gear train illustrate a combination of two existing technologies—as the result of which a new kinematic property is derived. As is clear from the graph, the speed of the wheel 2 is close to zero during the interval of the rotation of wheel 1.

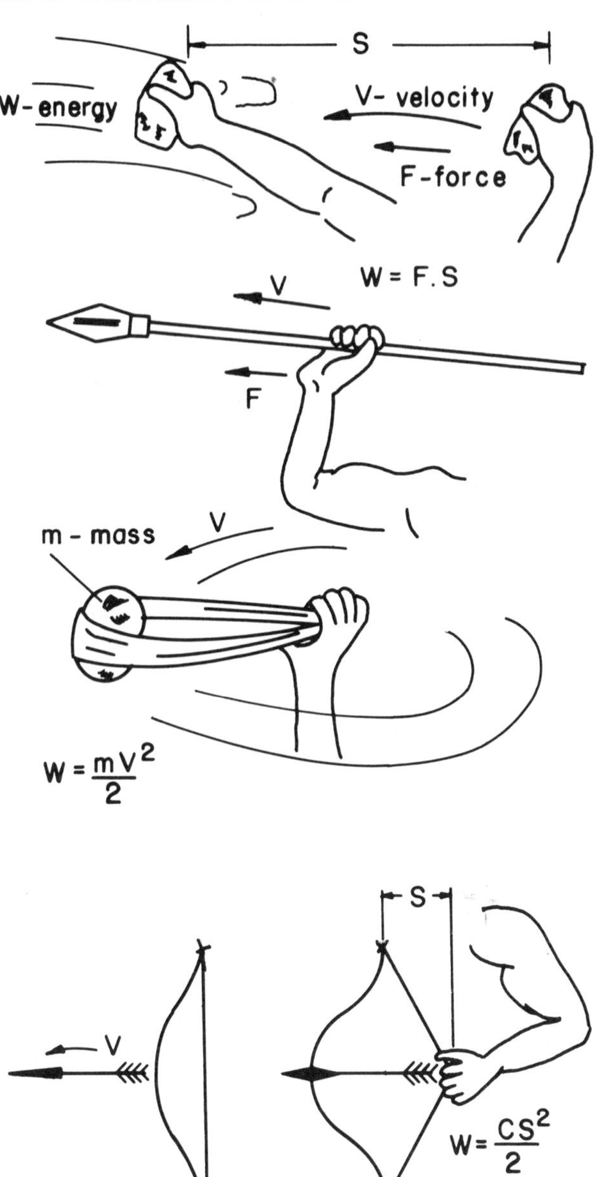

Figure 1-3 What is the difference between using a bow or sling and throwing a stone or a spear? The warrior accumulates the energy necessary for "throwing" the arrow by stretching the bow string (or rotating the sling). A new concept is thus achieved by separating the weapon from the hands of the warrior.

accumulates the energy necessary for "throwing" the arrow by stretching the bowstring and bending the bow. This is the first stage of the action. The second stage begins after the hunter has aimed: the hunter frees the bowstring and thus shoots the arrow. The same separation of actions can be seen in the use of the sling. The first energy-accumulating stage involves the rotating of the sling, and the second stage comprises freeing the sling and allowing a stone to fly in the desired direction. The first stage is manual, but the second is automated. The duration of the first stage depends on the capacity P, of the user, who has to attain a certain amount of potential energy W. Thus the time T is

$$T = \frac{W}{P}$$

The capacity of the device does not depend on the user, but the user's power can be used to actuate the device or tool. The efficiency of solutions of this type depends largely on the skill and personal attributes of their operators. The accumulated energy W is transformed into kinetic energy during a definite interval T_o, which depends on the bow's capacity P_o. *Thus* $T_o = W/P_o$. The value P_o is usually bigger than P, and therefore $T_o < T$.

Anybody who knows how to aim, has good eyes, and is trained to use a weapon can propel that weapon over a longer distance than would be possible in a completely manual case. The efficiency of the weapon thus is no longer a function of the hunter's physical properties as it was before the bow was invented. As a further illustration of the fourth principle, we can compare advances in technology with those achieved in sport. Today each centimeter, each fraction of a second, in sport requires great effort, sophisticated training, and a high level of scientific investigation. As in the technical field, after the main aims have been achieved, each additional improvement in an existing concept requires both a tremendous amount of research and large investments. We can depict this process in technology as well as in sport in a graphical manner. In the diagram shown in Figure 1-4, at points A_I and A_{II} it is no longer worth expending more effort on improvements; a new concept must be found because the "ceiling" of the concept has been reached.

To deepen our understanding of the fourth principle, let us look at another example—the history of the development of the bearing (Figure 1-5). The ancient bearing consisted of a wooden or stone shaft enveloped in a simple wooden or stone sleeve. Difficulties in the treatment of these materials resulted in low accuracy of the joint elements. With the use of metals, the quality of the bearing improved, especially after the invention of turning. A large contribution to and a new concept in the further development of the bearing was the use of lubricants, and by the choice of the right combination of materials, the limits of improvement were reached (Figure 1-5I).

The next step was therefore a new concept—the use of rolling elements in the bearing (i.e., the transition from sliding friction to rolling). With this improvement the losses in the bearing were decreased by ten or more times (Figure 1-5II). The first roller bearings were invented in 1862 for use in bicycles. The mass

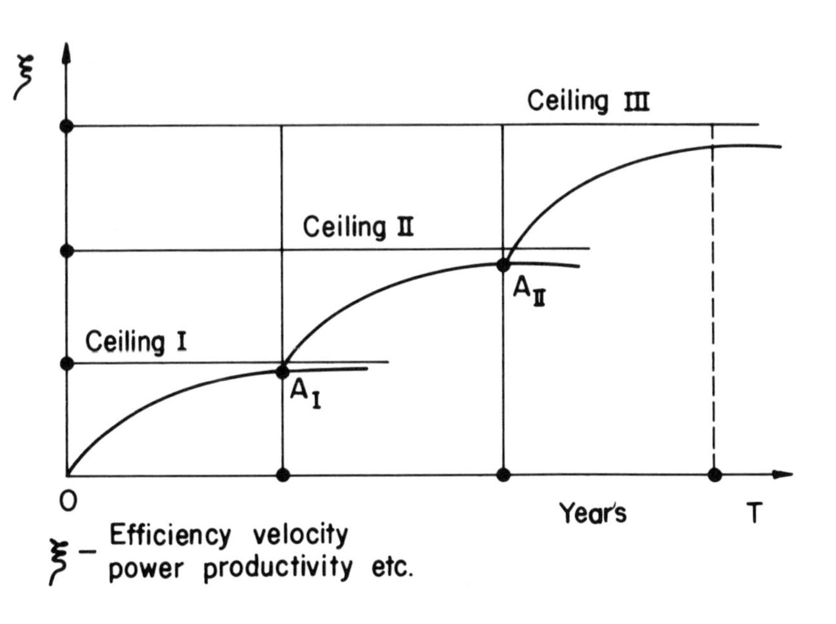

Figure 1-4 The interrupted nature of the development of technical concepts. At points A_I, A_{II}, . . . it is no longer worth expending effort on additional improvement of an existing concept. A new concept must be found where the "ceiling" of the concept has been reached.

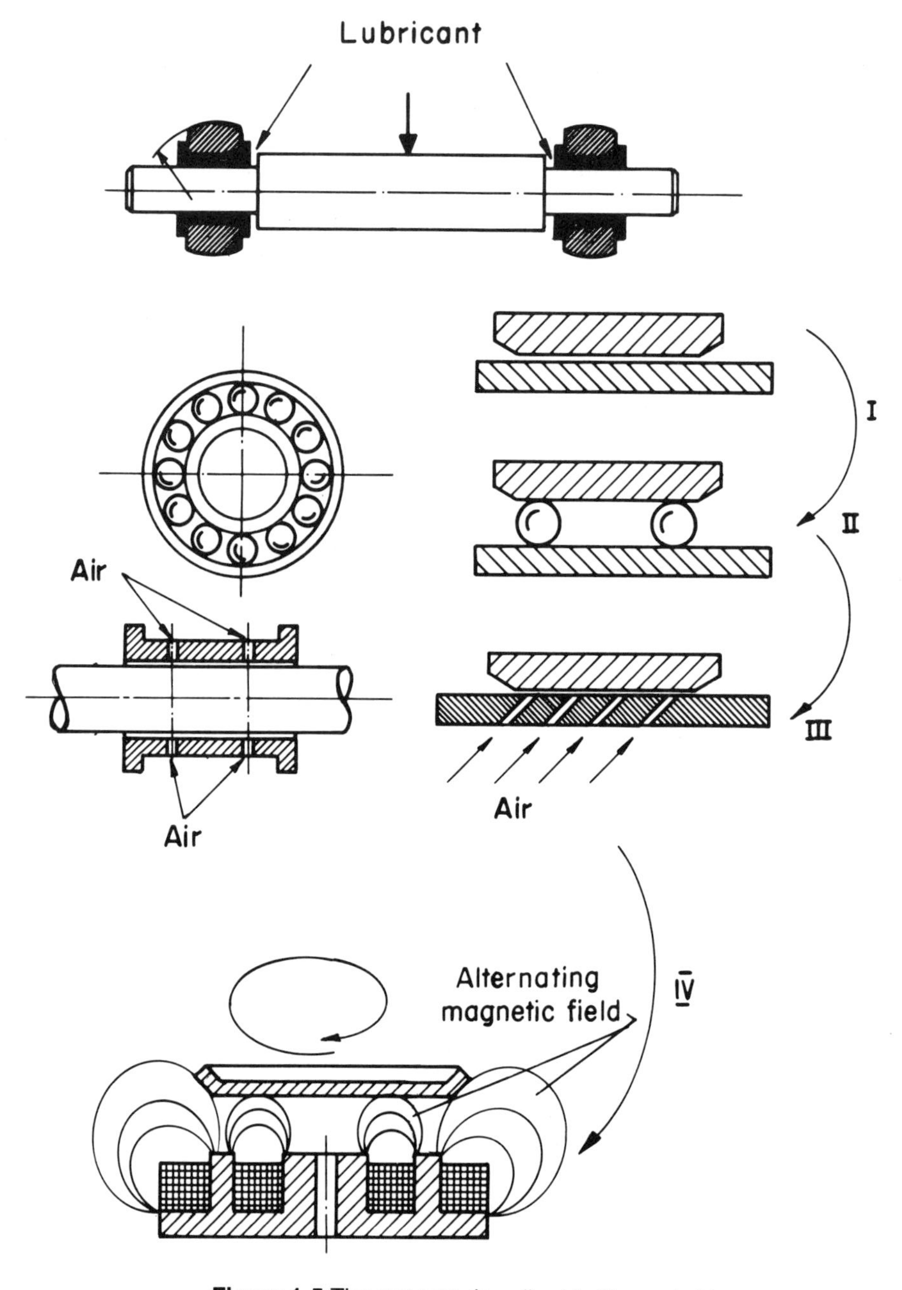

Figure 1-5 The concept described in Figure 1-4 is illustrated by the history of bearing development—from simple, lubricated friction I, to rolling friction II, to the air cushion III, and finally to electromagnetic suspension IV.

production of roller bearings and their applications in general machine manufacture began in 1898, and since then many kinds of roller bearings have been created. But for some technical requirements, these bearings are not smooth enough; that is, the resistance caused by friction due to the rolling movement is sometimes very large in comparison with the needs of a mechanism such as the gyroscope. To reduce this negative effect, a special "driven bearing" was proposed. The main concept behind this solution is illustrated in Figure 1-6. The external ring of the bearing is allowed to rotate, and the relative motion between the rings and rolling elements is thus decreased, as are the losses due to friction. We can see that this solution is very complicated. Moreover, although this solution improves one feature of the device, it introduces other troubles, such as vibration, noise, and loss of accuracy.

The next new idea was the use of pneumatic and hydraulic bearings (Figure 1-5III). In this type of bearing, the shaft "floats" in the sleeve on a thin layer of compressed air or liquid, and there is no direct contact between rigid surfaces. Friction is very much reduced, especially when compressed air is used. The rotation speed can be considerably increased without increasing the resistance, which is of great importance for gyroscopes and other

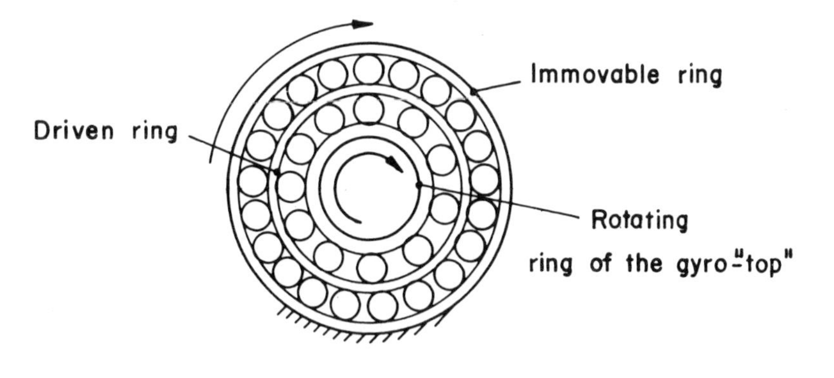

Figure 1-6 A double bearing for diminishing friction losses. The external ring of the bearing rotates so as to decrease the relative motion between the rings, thus reducing the losses due to friction.

devices that rotate at very fast speeds. It should be pointed out that the hydraulic bearing is similar to the lubricated slide bearing. The only difference lies in the fact that for the hydraulic bearing modern technology allows the elements to be fitted together with such a high degree of precision and such lubricant feeding systems to be produced that a thin liquid layer becomes realistic.

The next, and perhaps most revolutionary, idea was the suspension of the rotating body in a vacuum by means of a magnetic field. (Figure 1-5IV). In this case the speed of rotation reaches millions of revolutions per minute. This value is greater than the speed that most materials are able to resist because of the forces of inertia caused by the rotation.

Here we should remember that when one problem results in the appearance of a number of inventions, it is obvious that the most effective solution has still not been found, and we remain on the plateau of the exponent. A "revolutionary" solution is still required.

The Fifth Principle

At this point we can formulate the fifth principle of creativity.

V. A new solution can be found in the past history of the subject under consideration. The development of many products follows a spiral-like path. The same solutions return, but in significantly improved forms that impart to the product new qualities and properties.

The history of aviation is an outstanding example of this statement (Figure 1-7). In 1783 the Montgolfier brothers built the first fire-powered balloon and embarked on the first real flight known to mankind. This balloon was, in fact, the first flying aircraft. The concept underlying the use of this "lighter-than-air" device was borrowed from Archimedes' principle (in an application of Principle I). Difficulties in navigating this type of airborne craft resulted in the creation of a new device—the airship (dirigible)—which was a combination of a balloon and propelling equipment (Principle III). The first airship took off in 1852 after the "parent" balloon-type craft had been improved in many ways. New materials, such as aluminum were used for constructing the body, and

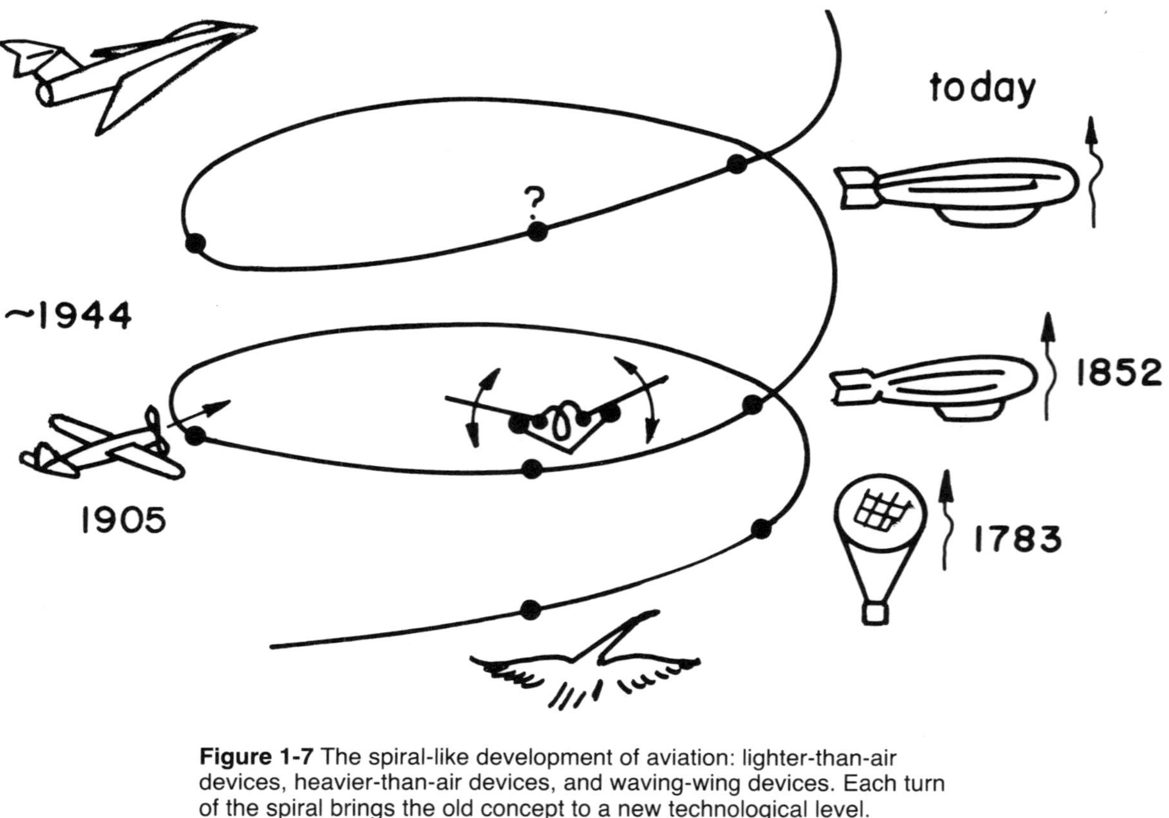

Figure 1-7 The spiral-like development of aviation: lighter-than-air devices, heavier-than-air devices, and waving-wing devices. Each turn of the spiral brings the old concept to a new technological level.

hydrogen replaced warm air (which had to be heated by a bonfire) as the "lighter-than-air" gas. The zeppelin disaster of 1937 put an end to this kind of flying for about 30 years. At present we are, however, witnessing the restoration of this concept, but with helium replacing the dangerous hydrogen gas. Thus the spiral that began in 1783 completed its first turn at the junction of the nineteenth and twentieth centuries with the creation of the airship and will make its second turn during our lifetimes.

A second line of development appeared on the spiral with the invention of the "heavier-than-air" aircraft by the Wright brothers. Orville and Wilbur Wright built their first airplane in 1903, and made their first practical flight in 1903.

There is a third line of development of flying devices (Figure 1-7) that should not be forgotten. It appears that a flying apparatus with moving wings (like those of a bird) is much more efficient than conventional aircraft, that is, the amount of the useful flying weight per unit of fuel weight or per unit of spent power is larger. Modern technology has enabled the construction of such a waving-wings model. This type of craft may well be the airplane of the future, or at least the small airplane of the future—it is difficult to imagine a jumbo jet waving its wings. Thus the third line on the spiral winds from the bird through Icarus, the son of Daedalus, to the experimental flying ornithopter.

Another excellent example of the spiral development of technical ideas is provided by a commonplace instrument—the razor. We suppose that in ancient times it was a variation of knife, but it is unlikely that a stone-made knife could afford much pleasure to the user. In more modern times, with the improvement of the properties of steel and the invention of stainless steel (and the development of shaving foams), the straight razor came into use. These instruments are still used by barbers today. Gillette then invented a practical, cheap, and convenient blade for the safety razor, which precluded the need to sharpen the blade. The manufacturers of razors subsequently produced various kinds of blades, razor handles, and auxiliary accessories for shaving.

In the middle of the twentieth century, the industry was struck by the invention and swift introduction to the market of electric and mechanical razors. A large number of patents were granted, differing in the mode of mechanical or electric drive, as well as

in the cutting method. It then took some 15 to 20 years for blade manufacturers to recover and to revive the concept of the blade. The new manual razors appearing on the market incorporated achievements in steel technology and the latest developments in plastic materials. The high quality of beard cutting combined with extremely high safety and "productivity" have rendered the "old fashioned" razor highly competitive with electric and mechanical models.

The turns of the spiral depend on the state of technology at the time in question. Thus it is obviously important to study the history of technical solutions and to check whether now is perhaps the time to apply some long forgotten ideas.

The Sixth Principle

Let us proceed to the next principle.

VI. This is the principle of inversion, that is, the exchange of positions, functions, or movements of the elements or links of the system under development. This principle may be applied at any level of consideration or on any "what" or "how" sheet. The following examples illustrate this principle. It has been known since the beginning of the nineteenth century that when an electric current flows through a conductor, a magnetic field is set up. In 1831 Faraday demonstrated the inverted idea that when a conductor moves through a magnetic field in a certain manner, an electric current will appear in the conductor. This idea is known as the law of electromagnetic induction.

An outstanding example of inversion is the IBM Selectric typewriter. The classic typewriter, which has been in existence for about 120 years, consists of a linkage mechanism system to drive each type bar regardless of the kind of machine (electric or manual), a paper roller mounted on a carriage, and a housing. The paper roller and the carriage constitute a relatively massive body, which has to accomplish rapid, accurate periodic movements during typing. Obviously the larger the paper sheets and the more copies required, the higher are the accelerations and declarations of the carriage and the stronger and more harmful are the impacts of the moving parts of the mechanism; the typing rate thus may

be restricted. Even electrically powered machines of classical design suffer from the same inherent problems as the strictly manual machines. Typewriter design thus reached the "top" of the exponential curve, and a revolutionary idea had to be found. The design of the IBM Selectric typewriter is an illustration of the inversion principle. In this machine the massive carriage is fixed and a light globe-shaped printing element is allowed to move along the paper. Thus the dimensions of the paper and the number of copies required do not influence the dynamics of the mechanism. A single typing head with a small mass can be moved faster and more accurately than the massive carriage. It should be noted that this concept became realistic only because of the development of electronics. It is difficult to imagine that such a solution could have been found 50 years ago.

The development of the gas turbine was also based on an inversion approach. The Laval steam turbine, invented in 1889, had an inherent balancing problem as a result of the high speed of rotation (about 30,000–42,000 r/m). Attempts to increase the accuracy and the strength of the shaft did not provide a solution to the problem. The answer lay in an inversion of these attempts: A flexible, thin shaft was introduced and its self-balancing property solved the problem. This improvement opened the way for the introduction of high-speed steam turbines, and later gas turbines, into industry.

And last, is not a rocket or jet-propelled missile an inverted gun? The recoil of a gun is utilized to propel a rocket, and the body of the rocket can be considered a moving gun tube. Later we consider a number of examples of inversion at different levels of complexity.

The principle of inversion does not always involve a change in a directly opposite manner, such as up–down, move–stop, or rigid–flexible. Let us look, for example, at the arc lamp. The original arc lamp consisted of two carbon electrodes between which an electric arc was maintained. Since it is difficult to provide a constant gap between the edges of the electrodes, special regulating devices had to be used to control this gap as well as the brightness of the light. To overcome this problem, Yablochkov, in 1876, changed the relative positioning of the electrodes: He placed them one next to each other with a layer of special insu-

lation, comprised of nonconducting clay, between them. In this setup the gap between the burning electrodes remains constant, the electrodes burn uniformly, and they shorten at the same rate. This invention made the arc lamp cheap enough to be used for many public needs, for instance, the illumination of Opera Square in Paris and many other streets.

The Seventh Principle

The inversion principle is a very effective tool in the creation of new concepts and the improvement of existing ones. But the efficiency of the drilling process, for instance, can be increased by multiplying the number of drill heads in a drilling instrument instead of seeking a new concept, such as punching or burning out the required holes. This example brings us to the next principle.

VII. The joining together of identical elements can result in a new effect—the principle of multiplication. One of the oldest examples of the application of this principle is Archimedes' tackle, which was an assemblage of pulleys. Another example may be found in the fifteenth century, in Leonardo da Vinci's weapon consisting of 33 gun tubes—three rows of tubes, each containing 11 tubes, arranged one above the other in such a way that the shooting and charging of each row was accomplished separately in a certain sequence. (Bear this in mind when considering Principle VIII.)

In 1884 an English engineer, Charles Algernon Parson, applied the principle of multiplication to the development of the steam engine by creating a multistage turbine. With this device the utilization of steam energy was improved. A more modern example is the type of drilling machine used in mass production. These tools are equipped with a number of drills located in positions corresponding to those of the holes required in the part being processed.

A note of caution must be introduced here. Care must be taken in applying this principle (as with Principle II): the number of components cannot be increased infinitely. It is ridiculous to put together, say, 100 gun tubes or 1000 drill heads. However, when

used properly, the multiplicative principle improves productivity. For instance, six gun tubes assembled on a revolving drum create the possibility of increasing the shooting efficiency six times. At any moment each of the six gun tubes is undergoing a different action—charging, closing, shooting, and so on. For each gun tube, the shooting pace is the same as that of a single-barrel gun, but as a result of the combined action, the system shoots six times faster.

The Eighth Principle

The eighth principle is a logical extension of the foregoing principle.

VIII. Simultaneous execution of a number of actions of a device can enhance its productivity and result in the achievement of new effects. The general case for this principle can be represented by the diagram shown in Figure 1-8. A number of actions, say three, are carried out simultaneously, and despite the relative slowness of the process, the final output rate is three times faster. If we suppose that mechanism B carries out the final operation of some process that takes time T and consists of three actions A, B, and C, then a combination of three machines I, II, and III will manufacture the final product, within a time interval t, where $T = 3t$. When the number of separate actions exceeds three, the advantage of simultaneous execution is even more important. For example, a combination of this principle and that previously described is used in the design of 4 – , 6 – , 8 – , and 12 – cylinder (and more) internal combustion engines. The pace of the working strokes depends on the number of cylinders. As the number of cylinders is increased, the power increases, and the rotation becomes smoother.

It is obvious that the use of this eighth principle is based on the use of semi- or completely automatic solutions. For example, the crossbow machine gun unit designed by Leonardo da Vinci in about 1500 illustrates the fact that this principle cannot be applied to nonautomatic systems: An archer was suspended independently and aimed and fired bows charged by 20 big men on the outside of a wheel, which was responsible for activating the system. The bow charging force was estimated at 120,000 pounds.

The Ninth Principle

Automatic machine tools are often based on principles of multiplication as well as on the ninth principle.

IX. Automation (electronics, pneumatics, hydraulics, mechanics) produces new effects and improves the product, device, or tool. The effects of automation are a result of its "tirelessness," its smaller "response time," its independent manner of action, and its ability to carry out a task under conditions in which a human being cannot function.

What are the reasons for the difficulties in product improvement as the advancement approaches the top of the curve? What are the restrictions that make each subsequent step in the devel-

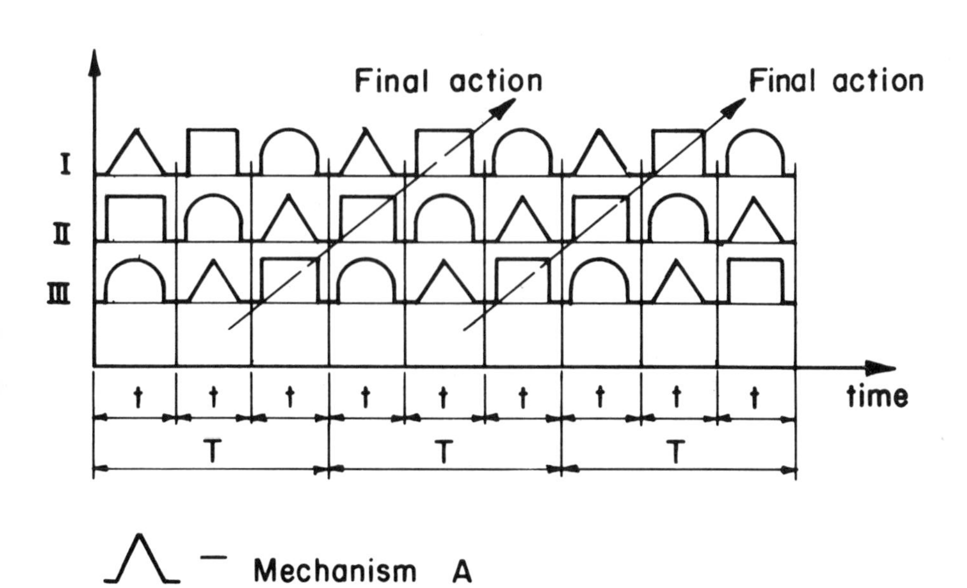

Figure 1-8 The simultaneous execution of a number of actions by a single device can enhance its productivity and result in the achievement of new effects.

opment of any technical concept less effective? The answers to these questions are not simple, but they may be said to revolve about three basic ideas:

1. The dynamics of mechanical, electrical, thermal, and chemical processes.
2. The properties of materials and substances.
3. The level of technology at our disposal.

For example, the IBM Selectric typewriter provided a revolutionary solution to the dynamically overloaded conventional typewriter. Similarly, the Otto internal-combustion engine was "replaced" by the diesel engine when the thermodynamics of the Otto engine had become "exhausted." The maximal efficiency of the Otto engine is about 35 percent while that of the diesel engine is 42–45 percent. An excellent example of the role played by technology may be found in the lagging behind of the Soviet Union in the "moon race" of the 1960s and 1970s. The low thermal resistance of the metals used for constructing rocket jets forced the designers of one Russian spaceship to propel the craft with 20 "low-temperature" engines (multiplication principle), whereas the Americans were able to drive their Saturn rocket with as few as four engines. The Soviet program was thus hampered by larger fuel requirements and heavier missiles. In addition, the lag in electronics technology reduced both the maneuverability of the Russian spaceship and its ability to join up with other craft in space.

What are the difficulties facing an engineer, a designer, or an investigator seeking a new solution or an effective concept? What are the main factors restricting such thinking? We consider them to be:

1. The absence of information in particular fields of knowledge.
2. The level of technology at the time under consideration.
3. "Psychological inertia."

The first two points do not require an explanation, but let us dwell briefly on the third point. The phrase "psychological inertia" is used to describe a common phenomenon from which all of us suffer, whether or not we are creators, investigators, developers,

or designers. The phenomenon may also be termed "conservative thinking." Let us take an example from ancient Rome. The Romans built hundreds of miles of aqueducts from force of habit because they knew that water can flow from a high point to a lower one. They could, however, have transported water in any direction by means of pipelines, since they knew how to manufacture pipes and they could easily have discovered the principle of syphoning.

This inertia thus may be described as the ever-constant striving to transfer the shape, concept, or structure of an old solution to a new one. The principle of copying solutions is generally useful, but in the particular case of transferring elements from an old concept to a new one, this principle becomes useless, and may even be harmful. We should be very careful to avoid this conservative thinking, this psychological inertia.

Three additional simple examples are appropriate here. The first steamships were propelled by paddle wheels partly because force of habit dictated that moving objects should have wheels. The first self-driven automobiles were carriage shaped (someone even fixed a sculpture of a horse to the front of these cars). One of the first electric motors was built like a steam engine: Two coils (like cylinders) pulled the armature, and a crank mechanism transformed the linear motion of the armature into rotation of the flywheel; a special switch mechanism (like a steam valve) controlled the coils.

Let us review the principles of creativity once more, this time as applied to the history of the development of gear transmission. We thus start with a brief study of the evolution of the art of gearing, which began about 5000 years ago. Unfortunately we do not know when the ancient "inventors" first proposed the idea of the wheel in general, but the inventors of the first geared wheels obviously borrowed the shape of the rotating wheel to utilize the remarkable properties of equidistant points rotating about a constant pin (Principle I). The principle of the engagement of two or more bodies for purposes of movement was also borrowed by these inventors of gears, who were familiar with this principle from their daily activities and skills. The teeth of the gears of that time had the shape the manufacturer was able to produce; that is, the level of technology dictated the structure of gearing. However, even in those centuries parallel, perpendicular, and crossing shafts

could be driven by means of gears. At this stage we approach the top of the first exponential* curve (see Figure 1-4, point A_I).

In the fifteenth century, the German scholar and mathematician Nicholas of Cusa (1451) began to study the application of cycloidal teeth to gearing, and in 1525 Albrecht Durer (of Germany) discovered epicyclic transmissions. These advances were followed by the development of involute gearing in 1694 by Philip de la Hire (France) and in 1754 by Leonard Euler (Russia) (Principle IV). At this time many mechanical clocks and watches were being built, and gear wheels of all dimensions were used for clocks from the size of Big Ben to tiny wrist, necklace, and pocket watches (Principle II). The demand for these types of mechanisms stimulated the art of gearing and the development of the gear industry. For example, the involved mechanisms needed for driving "puppet shows" led to the development of gear chains and complicated interlacings of shafts, wheels, and spindles (Principle V). The calculated cycloidal or involute teeth profile brought new qualities to gearing, that is, the motion became uniform, higher speeds became possible, and the ratio became constant. In 1852 Edward Sang (Scotland) wrote his general theory of gearing and teeth generating, and the top of the second exponential curve was thus reached. Setting aside many minor although important details, we notice that during that period the main struggle was directed toward increasing the durability of the teeth. In general their ability to withstand bending was much greater than their contact strength, since two convex profiles cannot provide good contact conditions. There is only one case in which involute teeth do allow good contact, that of inner engagement. Because of their structural disadvantages, these kinds of gears are used relatively infrequently. Increased dimensions, better materials, thermal treatment, and spiral and "herring-bone" (Principle VII—two helical gears together) teeth have all contributed to narrowing the gap between the bending and contact capacity of gears (Figure 1-4. point A_{II}). Although these factors improved gearing, they did not close the gap completely (Principle II).

The next exponential section was born with the introduction, by Wildhaber (of the United States) in 1921 and Novikov (of the U.S.S.R.) in 1955, of a fundamental change in the implementation of the basic gearing rules—that during wheel rotation, the pressure

* The curve may be described as being exponential because its shape may be approximated by an expression such as $A(l - e^{-at})$.

line must pass through the instantaneous center of the mechanism, and that the profile contact point that lies on the pressure line must slide along this line (as in involute engagement) or along some other curve (as in cycloidal engagement). In both cases the pressure line and the trajectories of the contact points are situated in planes perpendicular to the rotation axes. The innovation of Wildhaber and Novikov lies in the fact that the basic rule can be implemented in other planes, for instance, in a plane tangential to the pitch cylinders. The contact point slides along the teeth (which have to be spiral). In applications of this innovation, the profiles of the pinion teeth are usually convex while those of the wheel teeth are concave, both being helicoidal. Wildhaber and Novikov thus overcame the psychological inertia of traditional thinking and obtained a new effect (Principle IV). The contact capability of this arc-shaped convex–concave profile is about twice as high as that of the involute wheels. At this stage bending begins to become the major limitation. The matching of the concave and convex profiles of circular arc gearing is shown in Figure 1-9a and 1-9b.

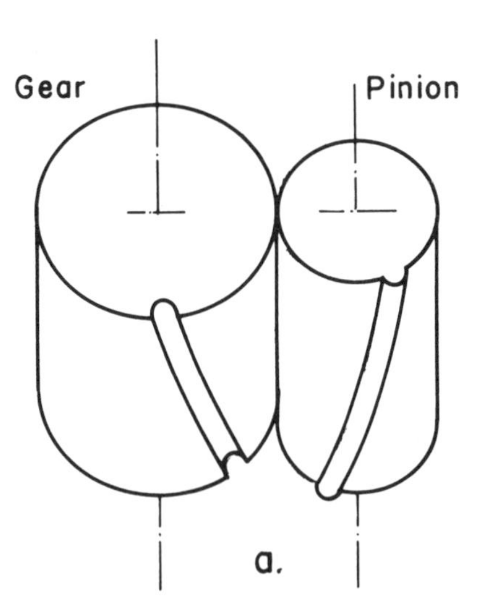

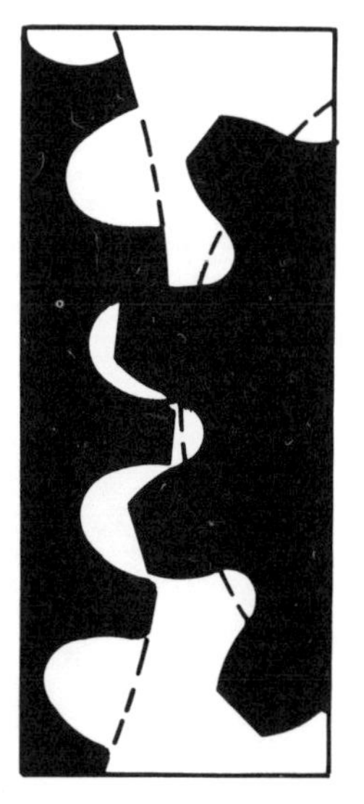

Figure 1-9 The gear-and-pinion circular-arc wheels shown (a) are provided with concave and convex teeth, respectively (b). This gearing concept is novel in that the line of action lies in the plane perpendicular to the plane of rotation of the wheels.

There is still room for improvement of these circular arc gears, particularly by finding solutions to problems of grinding, measuring, and other manufacturing difficulties. We are now at the beginning of the third exponential; point A_{II} (Figure 1-4) is behind us.

Chapter **2**

COMPLEXITY OF PROBLEMS

The technical problems we have to solve today are varied in their subject matter and differ in their levels of complexity. Each level of complexity requires its own tools and means for problem solving, and, obviously, the intellectual power required for each level also differs.

ALTSHULLER'S METHOD

Altshuller[1] has proposed a method of classifying problems according to their complexity, which is based on the following criteria:

1. The way in which the problem is chosen.
2. The way in which the concept of the solution is chosen.
3. The way in which the necessary information is gathered.
4. The way in which the principle of the solution is chosen.
5. The way in which the design is carried out.

Table 2-1 shows how the criteria change in accordance with the level of complexity. The level of complexity of a problem may be said to be the one corresponding to highest level into which any of the criteria fit. There are, of course, no "rigid" borders between the levels. But in typical cases there is no doubt as to

Table 2-1
Level of Complexity According to the Criteria of Altshuller

Level of Complexity	*Method of Choice of Problem*	*Method of Choice of Concept*	*Means of Gathering Information*	*Method of Choice of Solution*	*Method of Design (Creative Stage)*
I	Trivial problem	Use of existing concept	Common knowledge	Trivial solution	Trivial design
II	A problem selected from several problems	A concept selected from several possibilities	Information gathered from several sources	A solution picked from several possibilities	The design selected from several choices
III	The initial problem is changed	The concept is changed in accordance with the problem	The information is treated according to the changed problem	The solution is changed	The initial design is changed
IV	A new problem is found	A new concept is found	New data relating to the problem are obtained	A new solution is found	A new design is used
V	A new domain of problems is discovered	A new method is discovered	New data relating to the new method and the problems are obtained	A new principle is discovered	New design principles are created

which level the problem under consideration belongs. We shall illustrate the concept of complexity levels by a number of examples.

THE LEVELS OF COMPLEXITY

The first complexity level includes, for instance, problems that can be formulated as follows:

Making something lighter.
Making something stronger.
Making something more accurate.
Emphasizing particular properties, such as heat transfer or heat insulation.
Combining certain simple functions.

The concepts employed to solve these problems are trivial and may include the use of alternative materials, such as plastics or special alloys for reinforcement; the use of special shapes or structures, such thicker parts, crossbars, ribs, openings, or special cross sections; or changes in the methods of manufacturing, such as precise casting and grinding.

The information used to solve these problems can be found in engineering textbooks, handbooks, and catalogs. This information often belongs in the "common knowledge" category, and no printed matter is needed to find the solution. Thus the solution and design do not require special creativity and are based on the utilization of information.

Let us look at three examples:

1. The problem is to prevent a door from knocking against a wall when the door is being opened. This problem is trivial enough to be included in the first level of complexity. An existing concept may be applied for its solution, that is, a mechanical support can be used to limit the movement of the door. No special information is needed to solve the problem. Thus the solution is also trivial, as is the design. The door had to be stopped and a support was used for this purpose.

2. The problem is to prevent the overheating of a transistor under an electric load. This problem is also trivial. The concept underlying the solution is the provision of good heat transfer from the transistor to the air. The necessary information can be found in any electronics handbook. The solution lies in the use of a heat radiator. The design is trivial: an aluminum body provided by ribs to increase the air contact area.

3. The problem is to provide a high degree of stiffness to a long shaft in a case in which no additional support or weight increase is permitted. The concept is mechanical. Any textbook dealing with the strength of materials will give the solution. The design involves the use of a hollow shaft constructed in such a way as to maximize the resistance moment of its cross-section and in this way to decrease the possible deflections under load.

The second complexity level can be illustrated by the following examples.

1. The problem is to produce a door that closes automatically after a person has passed through it. Here one first must formulate the conditions of closing. What will the speed be? Must it be uniform during the closing or is deceleration needed near the closing point (to avoid a strong stroke and noise)? What should the value of the closing force be? Must it change when a disturbance (entrance by a child or elderly person) occurs inside the doorway? How must the closing device behave during door opening? At this level obviously, one first must select the particular problem. Then one has to decide whether a mechanical, pneumatic, hydraulic, or electrical concept will be chosen for its solution. The sources of information will then correspond to the type of problem and the concept to be applied in its solution. Likewise, the solution itself and the design must correspond to the concept.

2. The problem is to create an automatic tablet dispenser that would ensure proper distribution of medicine for a sick person who has an active timetable during the day. When a person who leads an active life has to take three or more kinds of medicine in strict succession, at different times of the day, adherence to the medicine-taking schedule can become a serious problem.

Figure 2-1 shows the development of the design used to solve the second problem. The initial device had the form of a round box provided with a time scale and 24 sockets, 12 on each face (analogous to a watch). In the morning the patient would insert the pills into the relevant sockets, and during the day he would take them according to the sequence shown on the device. Experience showed that in practice 12 sockets were sufficient, since most patients do not take medicine at night. The second picture shows the addition of an electronic timer to the 12-socket device. This timer beeps at predetermined times to remind the patient to take the medicine. The number of sockets can thus be further reduced; the dosage time is monitored by an electronic device, rather than indicated by the location of the sockets. As shown in the third picture, a further reduction of size could be accomplished by separating the electronic timer from the tablet dispenser. At this stage "psychological inertia" had to be overcome before the round shape of the socket box could be changed. In the last picture, we see that the shape of the box is now right-angled, the number of sockets is reduced to six, and the volume of the box is efficiently utilized. Such a medicine dispenser can be conveniently kept in a pocket and will remind the patient to take the necessary medicine in the right sequence with no major interference with a busy day. Further improvements are, of course, possible.

We can see that, at least at the design stage, the designer has to choose the best of several designs. This case thus fits into the second complexity level.

Let us now look at two examples illustrating the third complexity level.

1. The problem is to find a way to fasten the parts to be ground to the table of a grinding machine where the parts are composed of non-magnetic materials. The widely used "electromagnetic table" is suitable only for steel and other iron alloys. What can be done when nonmagnetic material has to be treated? We must change the fastening concept. Obviously the trivial case in which the parts are provided by special claws that can be fastened by mechanical means is not universal enough. Here we see that the initial concept has been changed. For the purpose of finding a

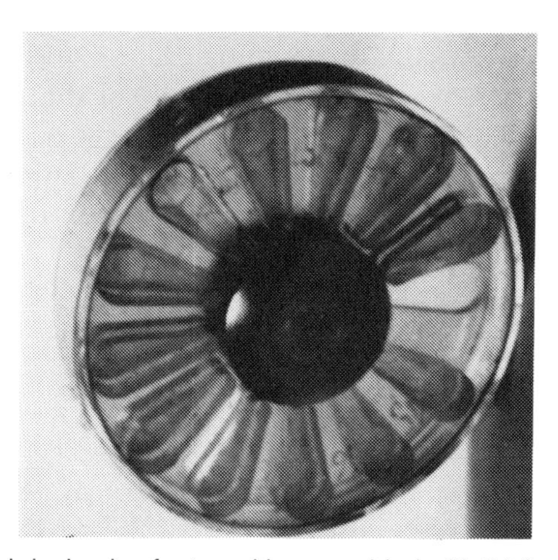

1. The initial design is of a round box provided with 24 (or 12) sockets, analogous to a watch.

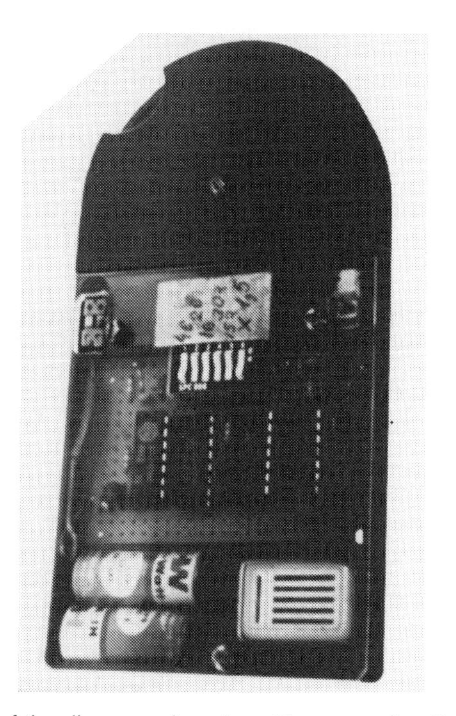

3. The size of the dispenser is reduced by separating the dispenser from the timer.

Figure 2-1 Four stages in the design of a tablet dispenser.

2. An electronic timer is inserted into the 12-socket box.

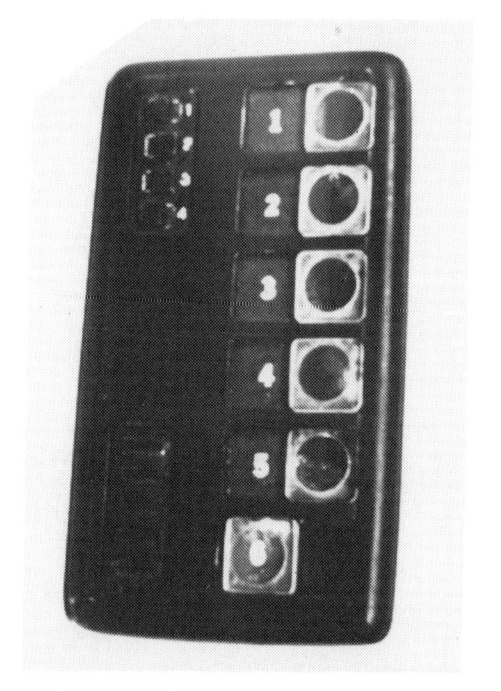

4. The size and shape of the device are changed so as fully to utilize the dispenser.

universal solution, a cooling device is mounted on the moving table of the grinding machine, and the part or blank is fastened to the table by allowing a liquid to freeze in such a way that it holds the part to the table.

2. The problem is to produce an electric starter for a high-powered three-phase asynchronous motor. For this type of motor, at the instant at which the rotor is stationary, the current is approximately twice as high as that under nominal rotation conditions. The conventional starter consists of a number of contactors (about four) actuated by special means that switch out the active resistances in the rotor circuit to limit the current (Figure 2-2). A new electric starter, which differs from the conventional starter, has been designed for this type of motor (Figure 2-3). The following section from the manufacturer's prospectus for the new starter explains its principles of action.

> An inductive winding, connected in series with the rotor circuit of a slip-ring motor, is wound on an iron tube which

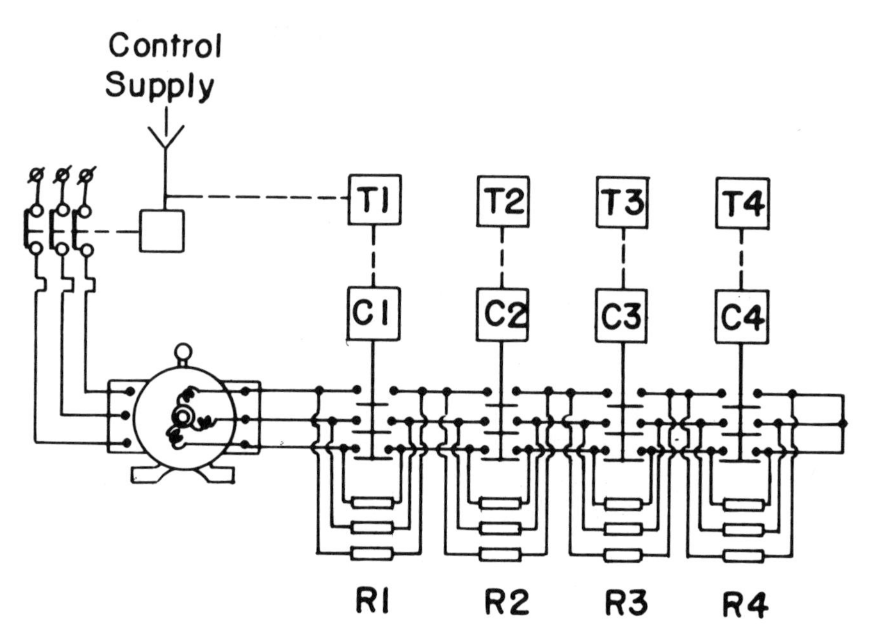

Figure 2-2 Principal layout of the electric circuit for conventional starting of a three-phase electric motor.

> acts as a short-circuited secondary of one turn. The iron losses caused by the eddy currents in the core result in an impedance in the rotor circuit which has a high resistive component. . . . This impedance is at its maximum at the instant that the mains voltage is applied and while the rotor is still stationary. The increase in the rotor speed during start-up decreases the iron losses which are a function of the rotor current frequency. This, in turn, results in a smooth, stepless reduction in the starter impedance as the rotor gains speed. In this manner, the rotor speed reaches approximately 92% of its synchronous speed, and at this point the rotor impedance is shorted out automatically. This operation is normally carried out at a predetermined number of seconds after the motor start button is pressed, but for special applications the shorting-out of the starter may be determined by the rotor current amplitude, speed or torque.

In comparing Figures 2-2 and 2-3, we can see that the new solution has considerably simplified the electric layout. This solution has

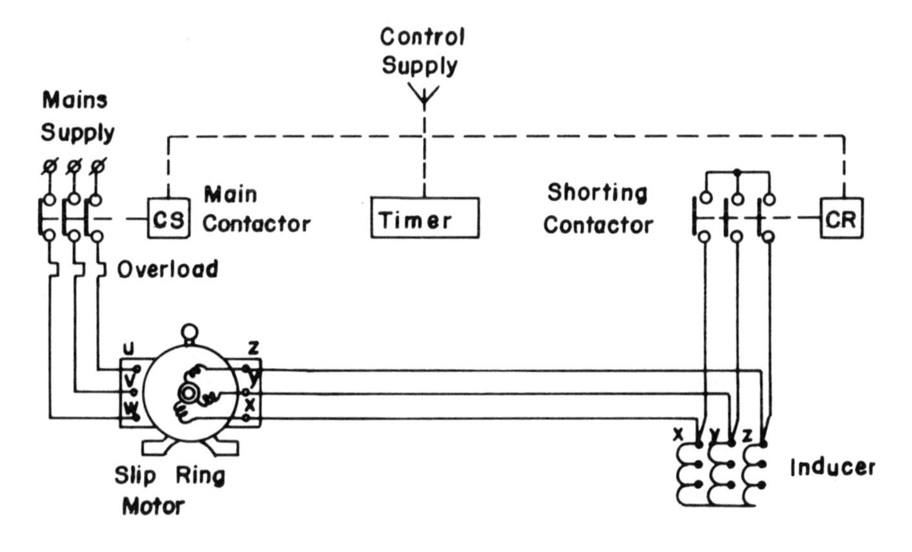

Figure 2-3 Principal layout of the electric circuit for starting a three-phase electric motor, where the starter is based on eddy-current effects.

the added advantages that it obviates stepped current increases (as takes place with resistor switching) and that it provides continuous, stepless acceleration.

Thus in the second problem the initial concept has been changed (a new approach is used, Principle IV), and the design has changed accordingly. Obviously we have a third-level technical solution. Note that we have not changed the main basic idea, that is, the dissipation of energy from the rotor in the form of heat in the starter. What is changed is the means by which the electricity is transformed into heat.

Let us now consider a problem illustrating the fourth complexity level.

The problem is to provide a plant with an optimal amount of irrigation water.

In solving this problem we note that there are many types of sprinklers that result in the soil being saturated with water. These irrigation methods entail a serious waste of water (evaporation from the air during sprinkling, evaporation from the open soil areas, etc.). A drip irrigation device was invented as an alternative solution. With this device water is supplied only to the roots of the plant via a pipe, that is, the water reaches its target in the shortest way with minimal losses. Thus a new problem was found (to irrigate the roots, not the soil);
a new concept was found (to create a dripping device);
a new solution and a new design were accomplished (a dripping system was created); and a method for calculating its parameters was developed.

As an example of the fifth complexity level, let us briefly consider holography. This technique is based on new electronic-optics methods whose introduction opened up a completely new domain of problems and applications. New data were gathered (and this gathering continues today) and a new principle for creating three-dimensional images, or holograms, was discovered. Special devices for producing the images, which can be viewed from different angles, were designed and constructed and are still being further developed.

Chapter 3

GENERATING SOLUTIONS

We have become familiar with some of the general principles we can use, consciously or unconsciously, in the development of new products. Let us put ourselves in the position of an inventor or developer and try to find what we can do to make the process of generating new ideas and new solutions more effective. It is easy to discuss the utilization of these principles and to illustrate them with examples, but it is much more complicated to solve a new problem when we do not know exactly at which "point" on the exponential we happen to be or whether a new solution can be found only by a leap in the technology. In other words, the process of seeking new solutions and generating new ideas is a system of overcoming contradictions. The point of the process is to find out what the contradictions are and to be clever enough to know how to resolve them. For instance, the contradiction that the Montgolfier brothers had to overcome was the fact that a human being is heavy and does not have the tools to fly. They followed (as was discussed earlier) the path proposed by Archimedes and built a device that had an average density less than that of air. The Wright brothers, on the other hand, chose an aerodynamic solution to overcome the same contradiction.

At this stage it would be useful to study some methods for the seeking of new ideas. Because thinking itself is under consideration, the methods must be psychological. It is useful to represent the solution-search process in a graphical manner. In Figure 3-1 we see the solutions plane. Each mark on this plane represents a

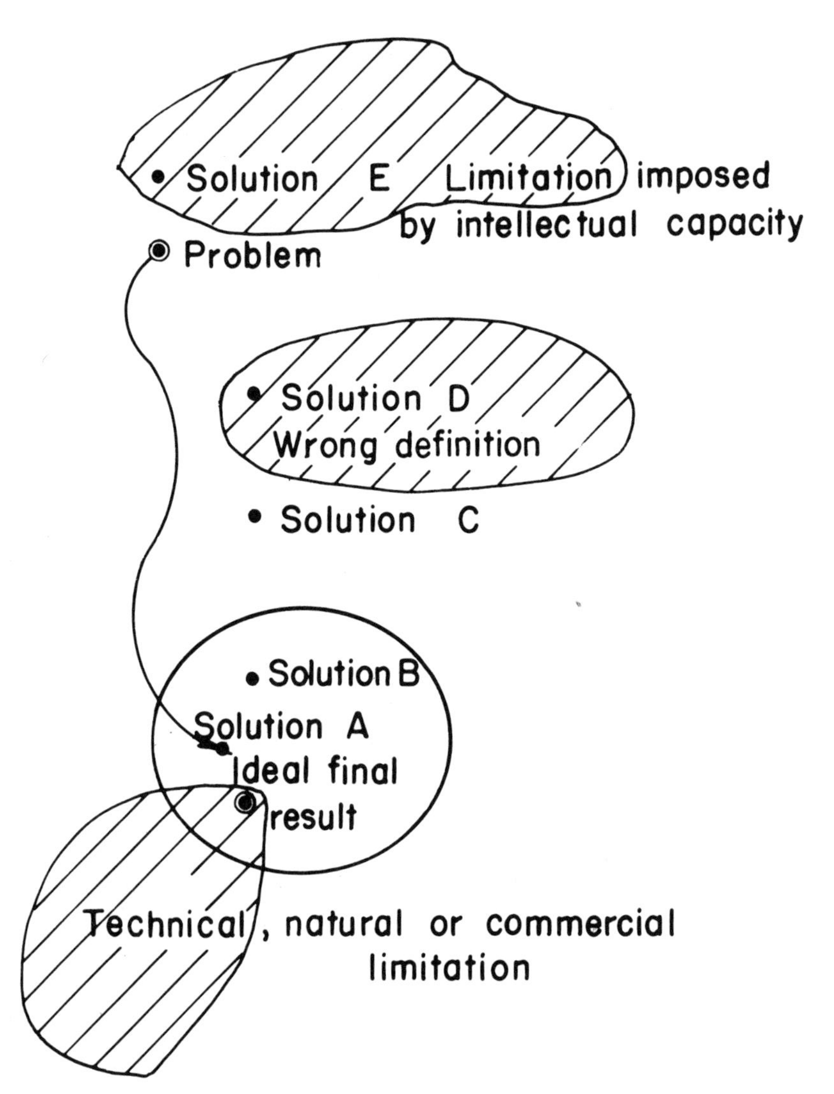

Figure 3-1 Graphic representation of the solution-search process. The shorter the trajectory of the search, the closer is the solution to the ideal and the higher is the efficiency of the design.

possible solution. The closer the marks, the closer are the solutions. The optimal case would be that in which an engineer travels from point to point—that is, from one solution to a better one—until the best solution is found or the time limit for the search is reached. But the restricted knowledge and mind of the designer will not permit following this optimal course, and an accidental solution is often chosen. In addition there are some restricted areas in the solutions plane. In Figure 3-1 we can see that there are three types of restricted areas:

1. Technical, natural, or commercial limitations. The example of the Soviet space program discussed earlier is an illustration of technical limitations.

2. Limitations imposed by false restrictions defined by the designer, by wrong formulations, by imprecise initial information, or by "psychological inertia." For example, it was thought that smooth cast-iron wheels rolling on rails would not be able to develop the friction essential to driving a train. The first steam engines, therefore, were equipped with teeth on the wheels (and rails) or "legs" to push them forward.

3. Limitations imposed by the intellectual capacity of the designer. For instance, a weak knowledge of electronics will lead a mechanical engineer to apply purely mechanical concepts to the solution of a problem, and some outstanding possibilities based on the use of electronics thus can be overlooked.

The following are methods used in the search for solutions.

1. Trial and error.
2. Brainstorming.
3. Analogy.
4. Empathy.
5. Inversion.
6. The morphological approach.
7. The systematic approach.

TRIAL AND ERROR

The most popular method used in the search for solutions is based on the knowledge and skill acquired by means of trial and error. The designer asks: "What will happen if I do this?" This is the simplest, and least effective, way of thinking. It was, however, the thinking and working style of Thomas Edison (1847–1931), who was the author of some 1200 patients in many different fields. His co-worker Nikola Tesla (1856–1948), the inventor of the three-phase synchronous electromotor, once said that if Edison had to find a needle in a haystack, he would be ready to disassemble the stack straw by straw. This was the way he found the "lighting material" for his electric bulb. The effectiveness of the method when employed by Edison can be explained in two ways: in terms of the staff that worked with him, and in terms of his own genius. We are, however, discussing methods for ordinary people. In Figure 3-2 we can see the leaps of a designer's thoughts, the "idle strokes" of the thinking process, until the solution dawns, as if by chance. This was the method used by the alchemists. Although they did a lot of work in vain, they managed to gather the large body of information that gave birth to modern chemistry. Many substances were discovered as "by-products." In 1674, for example, the alchemist Brand, in trying to recover from human hair a liquid for the transformation of silver into gold, discovered phosphorus. This trial-and-error method also resulted in the discovery of some important metal alloys. An outstanding example is duraluminum, a strong, hard, lightweight alloy widely used in aircraft construction, which was patented in 1910 in Germany by Alfred Wilm. He carried out about 10,000 different trials before he obtained an alloy with the properties he required. Of course, Wilm knew what he was seeking—which is an essential difference between this example and the previous one.

This manner of thinking is very widespread. All of us—children at play, dedicated technicians, lazy pupils, and even physicians—at one time or another espouse the slogan: **"Let us try it this way! Let us see what the response will be!"** Sometimes this is the only way to succeed. Relatively often, some very curious and amazing "by-products" are obtained. For example, a compass needle accidentally left near a wire carrying an electric current

led to the discovery of electromagnetism and the formulation of the laws of induction; a silver spoon forgotten on a metal plate stimulated the invention of the photograph; and a phosphorescent screen left near an active cathode tube gave birth to X-rays.

Trial and error is the natural response of human beings (and not only human beings) to difficult intellectual situations. It is the simplest way to gather skill and experience. The disadvantage of the "what if I do this" method is that the process of finding a solution to a problem involves moving from the beginning to the result, while the thought pattern moves in the direction of the vector of psychological inertia of the individual. Later we show that in many cases much more effective solutions exist in moving

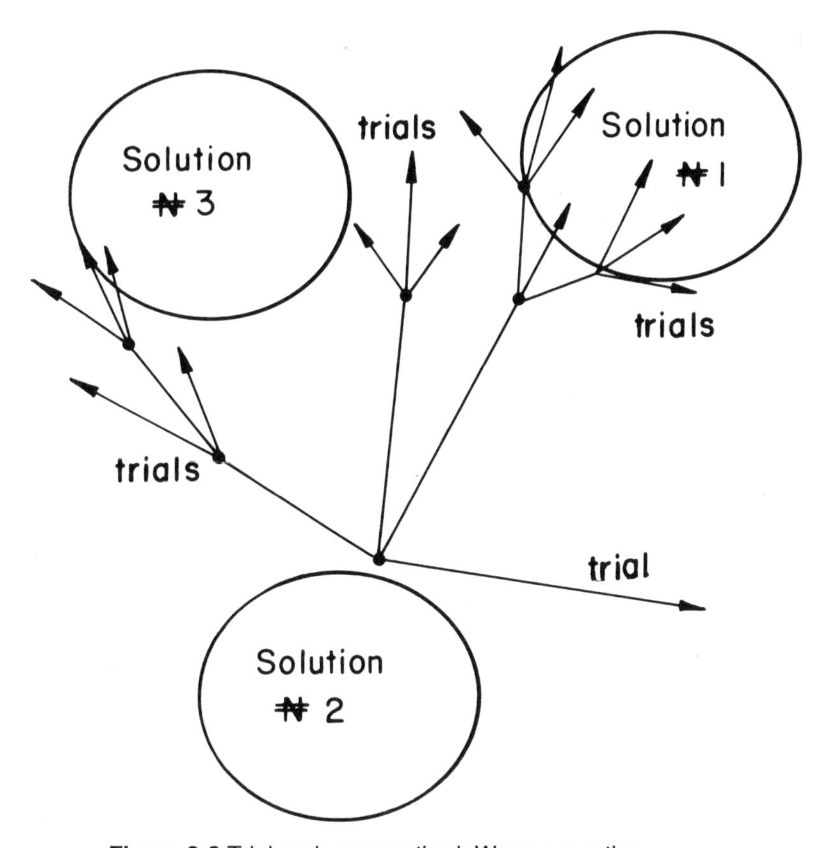

Figure 3-2 Trial-and-error method. We can see the leaps in the designer's thoughts—the "idle strokes"—until the solution dawns, as if by chance.

from the result to the beginning. All too often we are caught in the trap of straight thinking in which logic governs the mind, whereas creative, roundabout thinking requires logic to serve the mind.

BRAINSTORMING

The brainstorming method, which was proposed in 1953 by a U.S. psychologist, A.F. Osborn, is an improvement over the trial-and-error method. The novelty of this proposal lies in two main innovations:

1. A group of people, as opposed to a single individual, take part in the process of seeking a solution to a problem; that is, many brains are activated simultaneously (Principle VII).
2. The people who generate the ideas or solutions are completely separated from those who analyze the gathered solutions (Principle VIII, many brains are working both simultaneously and in parallel).

Note: Since this proposal of Osborn is a method of improving the solution-making process, we feel that it may be classified in terms of Principles (VII and VIII).

With the brainstorming technique (Figure 3-3), the chances of covering the solutions plane uniformly are increased, since the vectors of psychological inertia of the participants are randomly directed. The time required for generating new ideas is shorter because the technique involves a number of brains working separately and almost independently. The main conditions and requirements for a brainstorming session are:

1. The brainstorming group should include about ten persons.
2. These people can (or sometimes must) be specialists in several fields.
3. Each participant can express any ideas briefly, no matter how ridiculous they may sound.
4. The time limit for each pronouncement is about 1 minute.

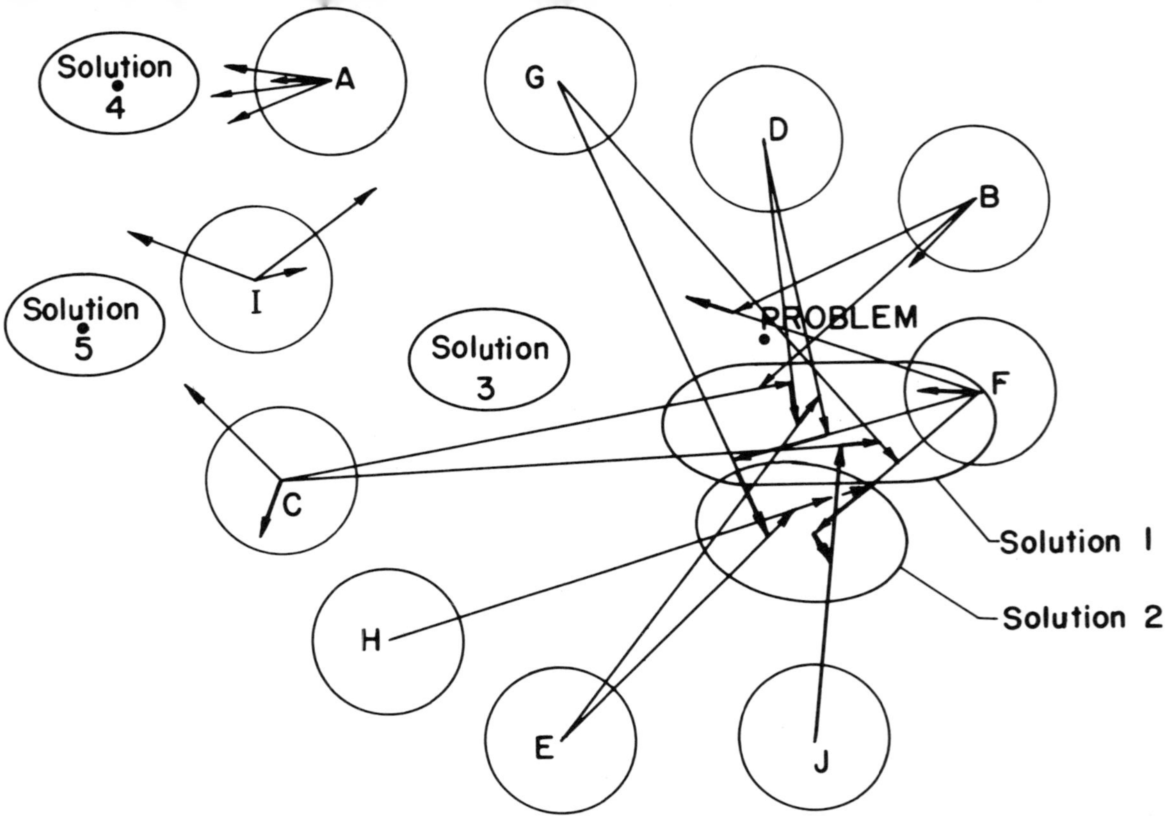

Figure 3-3 The brainstorming technique. The chances of covering the solution's plane uniformly are increased, since the vectors of the psychological inertia of the participants are randomly directed.

5. No criticisms or other remarks relating to the pronouncements are allowed.

6. The relationships among the participants must be free and friendly.

7. The session is to last about 40 minutes.

8. A group of analysts should record all statements and then rethink them carefully.

The main disadvantage of this method is that there is no control over the thinking process and a fruitful line of thought can be cut off by a participant's sudden suggestion.

Sometimes a lone creator can organize a brainstorming suggestion by speaking, for instance, into a tape recorder. In such a session, no doubts should be expressed, but only ideas without any restrictions, and with any amount of "ridiculousness."

An example of brainstorming session held by a group of about 15 students is presented here. The problem for which solutions were sought was how to steady a bicycle when its speed is low or even zero, as happens when it approaches traffic lights or is caught in a traffic jam.

Brainstorming Session

An idea for the design of a "steady-when-stationary" bicycle is required.

A: A collapsible support is needed.

B: A gyroscope would provide the required steadiness.

A: A collapsible wheel.

C: The gyroscope can be constructed in two planes—horizontal and vertical.

D: A gyroscope driven by a chain and treadle.

E: In this case the gyroscope can be placed inside the wheel—the rear wheel, for instance. No additional chain is needed.

C: Then a special coupling will be required.

F: It is possible to locate the gyroscope in a parallel plane, next to the wheel.

G: The gyroscope can be driven by an air turbine. An air compressor will be driven by the cyclist's feet.

E: A small roller can drive the gyroscope by friction from the wheel.

H: Then the system has to be arranged so that before braking the roller is pressed against the wheel, and when the riding speed decreases, it is freed.

B: Such a gyroscope can return its kinetic energy to the bicycle.

F: Such a gyroscope will, in general, smooth the running of the cycle.

H: It will promote steadiness during low-speed riding in traffic jams.

I: Two automatically controlled jets may keep the bicycle steady.

A: Instead of collapsible supports or auxiliary wheels, the whole frame can be lowered to the road.

F: An automatic device can press the roller to the wheel during acceleration and release it during deceleration.

C: I propose an air balloon to suspend the bicycle during its stops.

I: Or a propeller, like a helicopter, for balancing.

G: A horizontal gyroscope seems to be better; it will resist the bicycle turns to a lesser degree.

J: One merely needs a handle to switch the drive from the wheel to the gyro. Turning the pedals on the spot on will keep the balance.

C. A vertical gyroscope will not disturb the rider essentially, just as the wheels do not. It will help to incline the machine.

G: An electrically driven gyroscope. A bicycle provided by accumulators.

It was decided to use a gyroscope for the bicycle problem.

ANALOGY

Thinking in terms of analogies is a further development of Principle I. We can say that the copying of nature, of previously created devices, or of manual processes relates to "analogy thinking" just as arithmetic relates to algebra. (Note that to clarify the point, we had to use an analogy.) This method of thinking is very

useful in the conceptual stage of creating. When Rutherford propounded his concept of atomic structure, he described the atom as being like the solar system, although he knew that such a model could not exist. (Ten years later Niels Bohr improved this model by introducing some revolutionary explanations.) For decades this planetary analogy of atomic structure has been quoted in textbooks to help students understand the basic nature of the atom.

Let us consider some other well-known examples. There is an analogy between some electric and hydraulic phenomena; that is, some properties of direct electric current and the current of a liquid may be described analogously, such as potential difference and the resistance of wires and pipes. Another example is that of parametric vibrations in mechanical and electric circuits. Similarly, there is a direct analogy between the phenomenon of resonance in electric circuits and the vibrating of mechanical systems, and thus the mathematics used to describe the behavior of electric and mechanical oscillators is identical. This fact led to the development of analog computation. A wide range of electronic analog computers has been developed for many mechanical, hydraulic, and other investigations. In analog computations electric resistance plays the role of mechanical damping, inductance the role of inertial mass, and electric capacitance the role of stiffness. By knowing some special rules, one can translate mechanical values into electrical language, go through the computation, and then retranslate the results into terms of mechanics.

Another example of the application of analogies may be seen in the use of fluidics—the technology of employing the flow characteristics of a liquid or a gas to operate a control system. (The term fluidics is a combination of two words, fluid and logic.) This technique, which became commercially viable in the 1960s, is analogous to the electric relay control technique. The properties of fluidics render the technique unsuitable for some operations, and in such cases mechanical or electrical approaches have to be used.

All of these examples are examples of direct analogy, which is illustrated schematically in Figure 3-4. Here the domains of two phenomena are considered. We state that an analogy exists if there are parallel transformations from situation A_I to B_I and from A_{II} to B_{II}. Let us take domain I to be electricity and domain II to be

magnetism. The changes of electric current A_I to B_I and magnetic flow A_{II} to B_{II} can be described by identical analytic expressions. Thus an analogy exists. Analogy may now be used as a tool for discovering new facts. Let us consider an event in domain I—for instance, semiconductivity C_I to D_I—and ask ourselves whether an analogous situation, semiconductivity of the magnetic flow C_{II} to D_{II}, exists in domain II. Try to propose a layout that would conduct a magnetic field of only one polarity. In contrast to direct analogy, inverse analogy can be used; that is, if within a domain certain conditions A_I lead to a situation B_I, then by forcing the situation B_{II} in another domain, conditions A_{II} are obtained (see Fig. 3-5). For instance, when the junctions in a loop of dissimilar conductors are kept at different temperatures (A_I), an electromotive force (B_I) appears—the Seebeck effect. Inversely, when an electric current is maintained in a loop of dissimilar conductors (B_{II}), one junction of the conductors becomes cooler and the other one warmer (A_{II})—the Peltier effect.

A number of other examples may be quoted. Oersted discovered the connection between electric current and the behavior of the compass needle in 1826. Inverse analogy led Faraday to

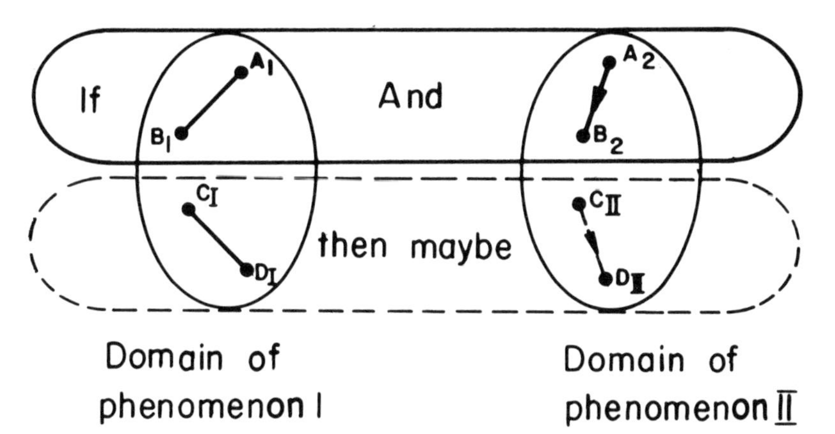

Figure 3-4 Direct analogy diagram. If two transformations A_I to B_I and A_2 to B_2 are alike in the domains of two different phenomena, then there is the possibility of finding a transformation C_{II} to D_{II} if C_I to D_I is known.

predict, and later to show (in 1831), the phenomenon of electromagnetic induction. The diffraction theory of a coherent light beam enabled the physicist Dennis Gabor to "invent" holography, which is an inverse phenomenon, in 1948. With the advent of lasers in the 1960s, a source of coherent light beams became a reality, and Dennis Gabor received the Nobel prize for physics in 1971.

As a practical example of the analogical approach, let us consider the creation of a remote-control irrigation system.

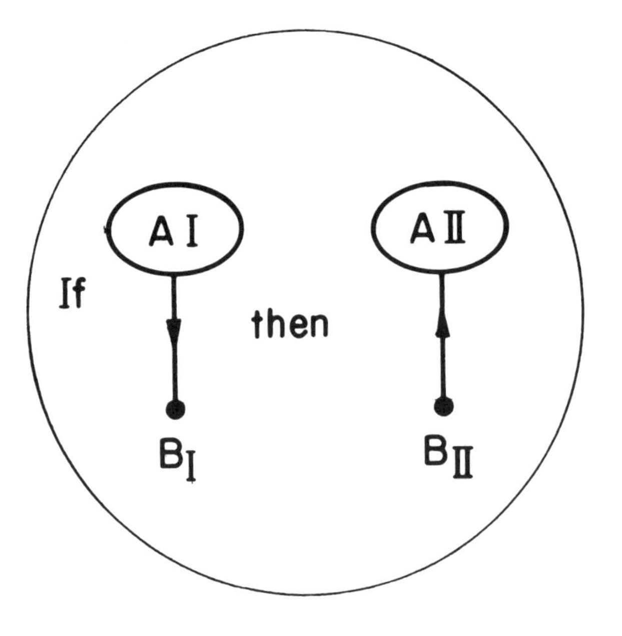

Figure 3-5 Inverse analogy diagram. If certain conditions A_I within a domain lead to a situation B_I, then by forcing the situation B_{II}, we can expect the appearance of conditions A_{II} in another domain.

Remote-Control Hydraulic System

Here we describe the analogous application of the electric frequency technique used in communications to a remote-control approach to the design of specific hydraulic systems. In some

hydraulic systems, it is necessary that the opening and closing of valves or taps be carried out by remote control, such as in the case in which the response time of the hydraulic servo devices must not be small and the distances between the control and serve units are significant. An automatic irrigation system is a good example of this type of remote-control hydraulic device. Our discussion will relate to this kind of system, regardless of whether or not the proposed approach can be used for other applications.

The overall layout of the type of system under consideration is given in Figure 3-6. A pipeline 1 connects the water (or other liquid) supply station 2 to the irrigation sprinklers S1–S6. The pipe has a number of branches, each of which is provided with a valve (or tap) VI–V6. These valves are controlled by control units C1–C6, which receive commands from the central command unit CO. A communication channel 3 connects the central command unit CO to the control units C1–C6. Energy sources E1–E6 are used for amplifying the commands arriving at units C1–C6. The power for actuating the valves is obtained from the water pressure by means of connections P1–P6. The communication channel 3 may be hydraulically, pneumatically, electrically, or radio controlled. Whichever means of control is used, an additional channel has to be constructed parallel to channel 3 to facilitate the remote control, and additional energy sources have to be installed. For instance, if radio control is chosen, the receivers have to be supplied with electricity from batteries or from an autonomous electric power generator (which may, for example, be wind driven). If a hydraulic system is chosen, a parallel pipe for transmitting control commands has to be installed. If an electrically operated variant is used, then at least one cable has to be laid along the system. The necessity for an additional control channel in pipes or wires makes the system cumbersome and expensive.

Let us now ask ourselves whether it is possible to design a simplified system without an additional control channel. Such a system is shown in Figure 3-7. In this new layout, there are two new elements (shown on the right-hand side) in addition to the elements shown in Figure 3-6 (on the left-hand side). These new elements are a frequency generator FG and a transmitter TR. When a command has to be sent to one of the valves or taps, the frequency generator produces pulses of a particular frequency that corresponds to the chosen valve and the required action (opening

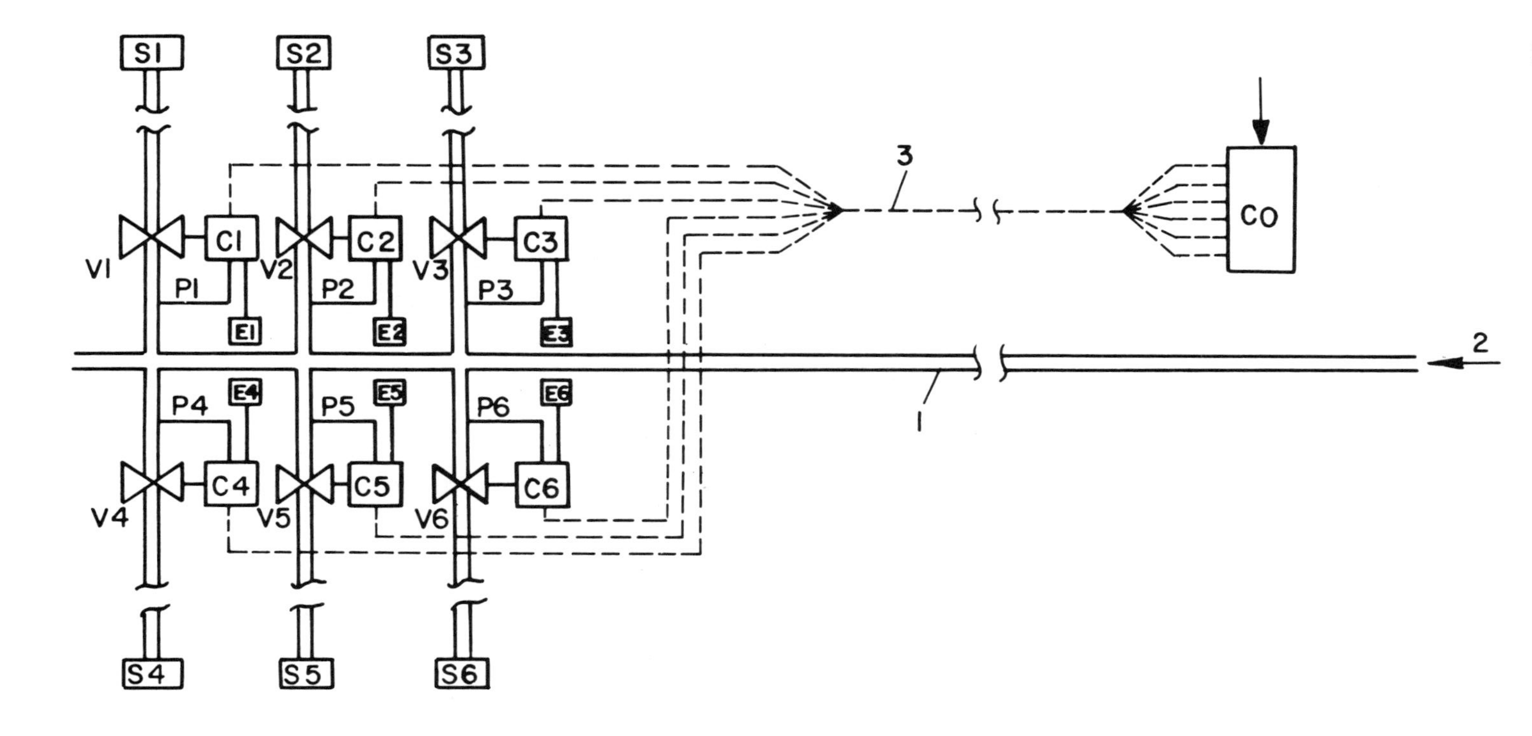

Figure 3-6 Overall layout of a conventional irrigation system. Here the water supply pipe and the control channel are separate.

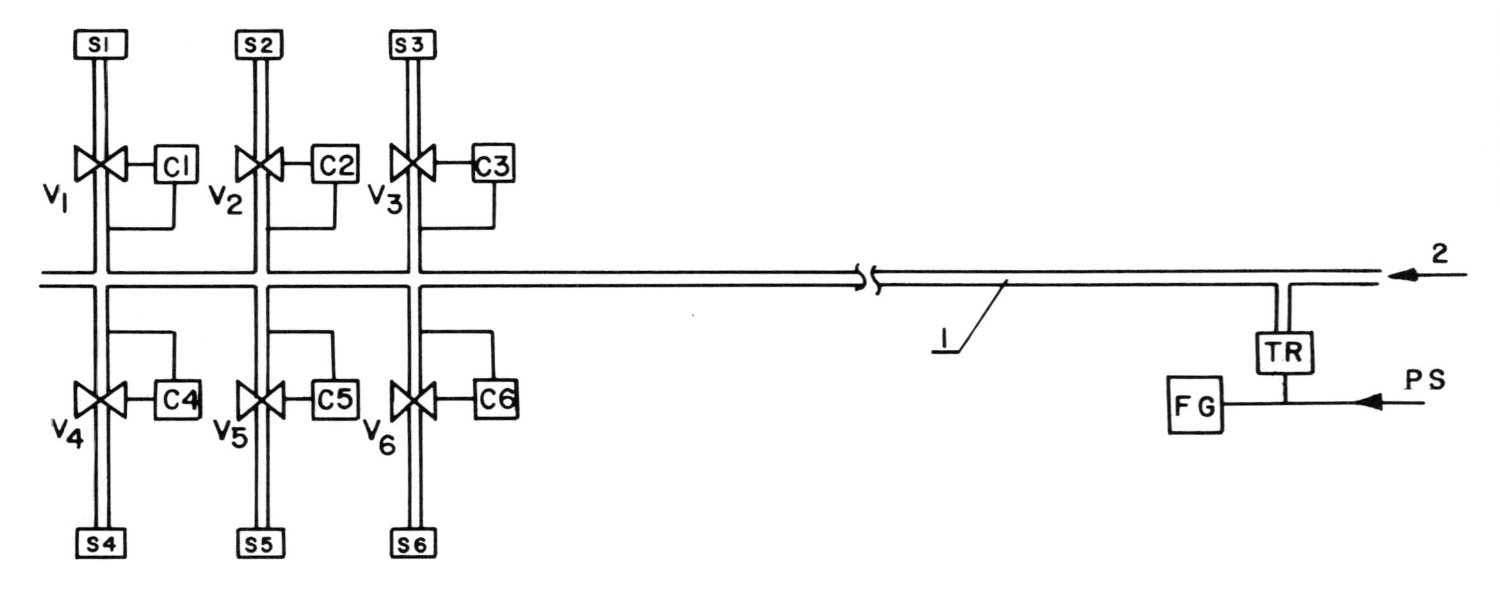

Figure 3-7 Overall layout of proposed irrigation system, in which the water-supply pipe also serves as the control channel.

or closing). The transmitter then transforms these pulses into pressure oscillations, which are set up in the pipe, so that the command is transmitted by pressure waves running along the pipe. The shape of the pressure waves is shown in Figure 3-8. The pressure P is the average pressure of the water supply. P_i describes the amplitudes of the alternating pressure. Each command requires a "package" of waves in a time interval if $\tau_1, \tau_2 \ldots \tau_i$ and a frequency ω_i given by

$$\omega_i \quad \frac{2\pi}{\tau_i}$$

Thus the instantaneous pressure p can be expressed as follows:

$$p = \bar{p} + p_i \sin \omega_i t$$

where t is the running time.

To examine this principle, we built an experimental device that would enable us to investigate the following points:

1. The influence of the length and shape of the pipe on the pressure-extinguishing process along the pipe.
2. The maximum possible number of different controllable elements that could be included in the system.
3. The nature and range of the disturbing events in this remote-control system.

To make it easier to understand our experimental device, let us look at Figure 3-9. Here one can see a transmitter consisting of a membrane 1, which is in contact with the liquid in the main pipe 2. The frequency generator FG actuates, for example, an electromagnetic valve EV, which connects (or disconnects) the membrane with the pressure source PS (in our case, we used compressed air). In this manner the pressure "package" can be added to the static pressure in pipe 2. The receiver consists of two membranes 3, which are connected by a rigid body 4. A bypass 5 is used to balance the static pressure in the system and to relieve the static pressure on the membranes. An orifice 6 is used to restrict the transfer of rapid pressure changes below the

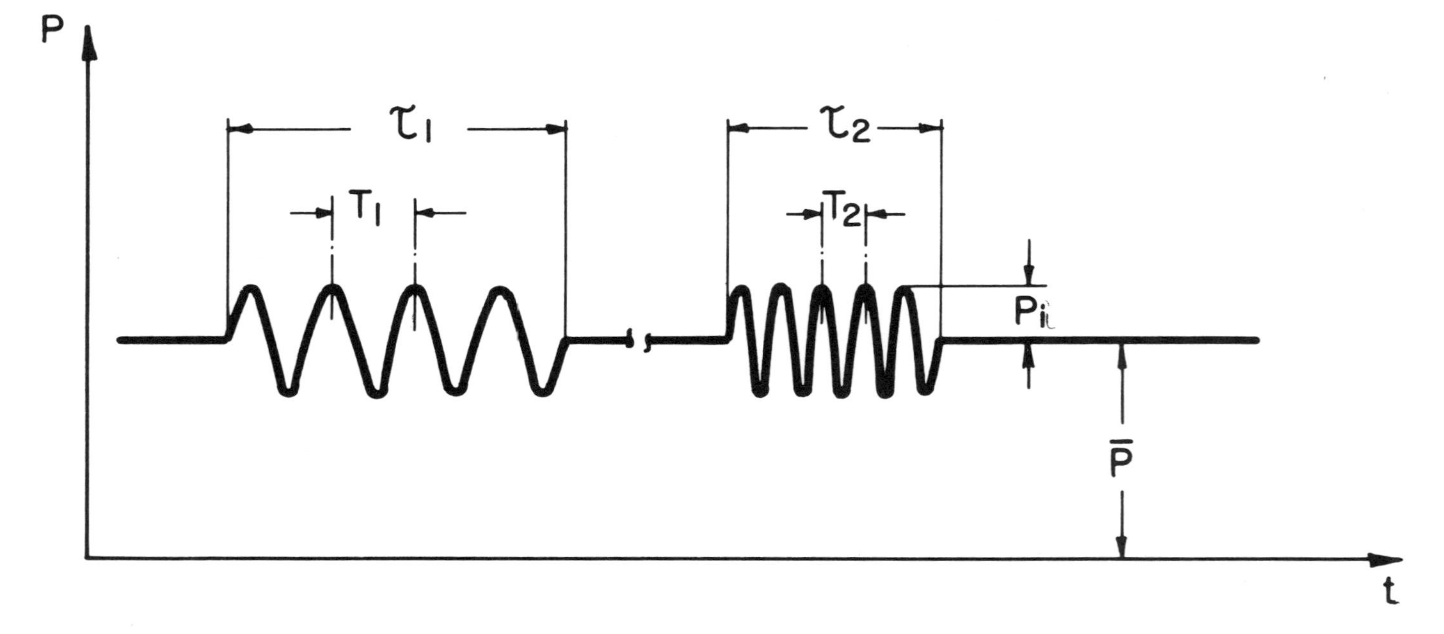

Figure 3-8 Shape of the pressure waves in the water pipe. The pressure comprises a constant pressure $\bar{P}$ (the water supply) and a variable pressure P_i (the control signal).

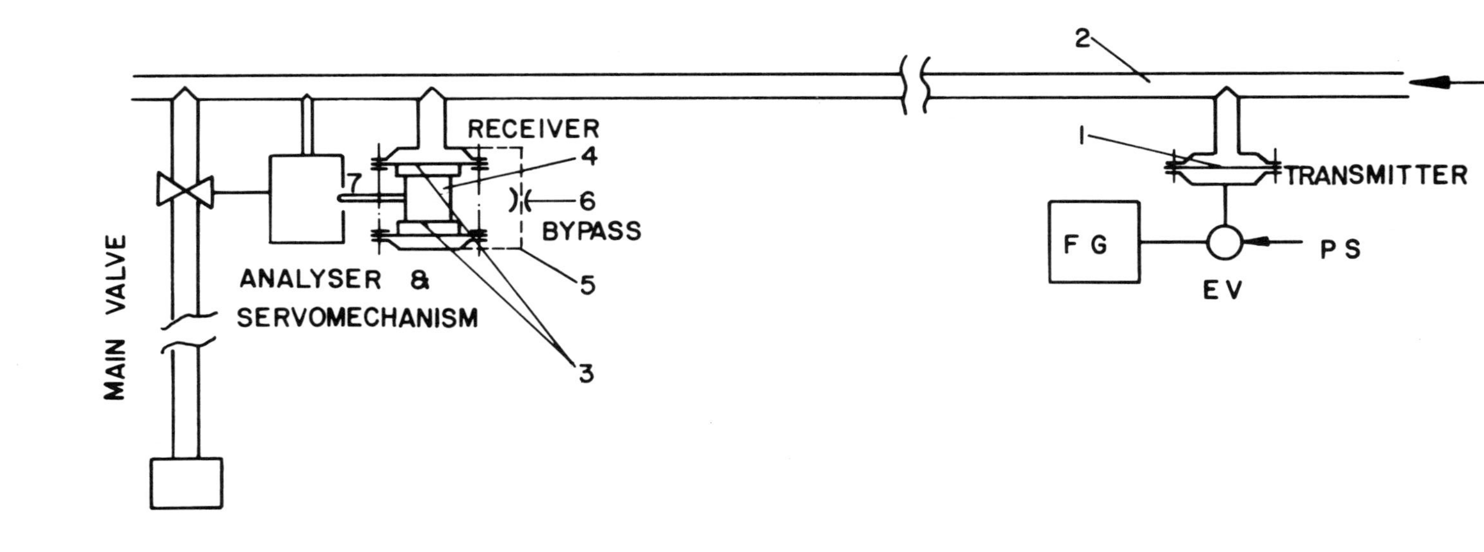

Figure 3-9 Layout of an experimental device consisting of a transmitter for generating P_i, a receiver for receiving the variable pressure P_i, an analyzer for separating frequencies of the pressure P_i, and a servomechanism for actuating the faucet.

lower membrane. As a result the body 4 oscillates with a frequency corresponding to the pressure changes in the pipe, and thus actuates the frequency analyzer (by means of a pin).

The mechanical design of the frequency analyzer is illustrated in Figure 3-10. From the figure we can see that the frequency analyzer consists of a shaft 1 driven by a pin 2. The pin receives oscillations from the oscillating body 4 shown in Figure 3-9. The free-mounted pendulums 3 are connected to the shaft by springs 4. The moment of inertia of the pendulums and the stiffness of the springs determine the natural frequencies of the receiver. When conditions at which resonance occurs are approached, one of the pendulums will begin to oscillate with increasing amplitude.

In our system each pendulum controls a single action; that is, the opening or closing of one tap. Each tap or valve is provided

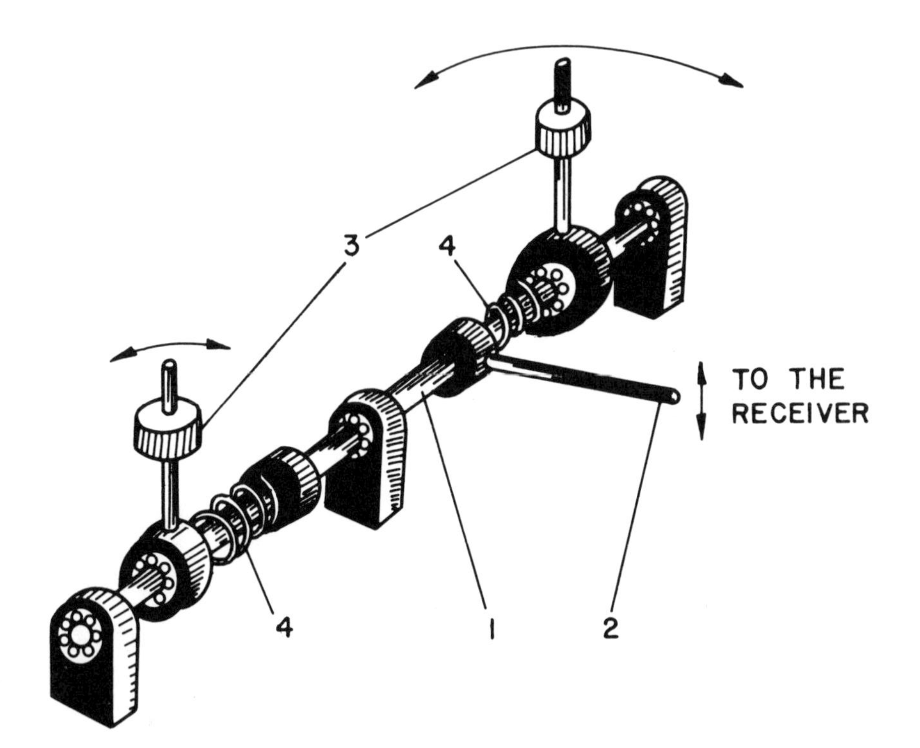

Figure 3-10 Mechanical design of an experimental water-pressure frequency analyzer.

with an identical device tuned to a different frequency. The oscillations are transformed into action of a tap or a valve by means of a hydraulic amplifier, which is shown in Figure 3-11. It consists of a housing 1 into which a piston 2, two plugs 3, and a system of channels are inserted. When the water pressure is transmitted to channels A, the piston 2 is balanced. If, while oscillating, a pendulum hits one of the levers 4, and in this way opens one of channels B, the balance of the piston becomes disturbed and it begins to move in the direction of the lower pressure. This forms a connection between channel C and either channel D or E; this last action transmits water pressure to the hydraulically actuated valves, causing them to open or close. Springs 5 press on levers

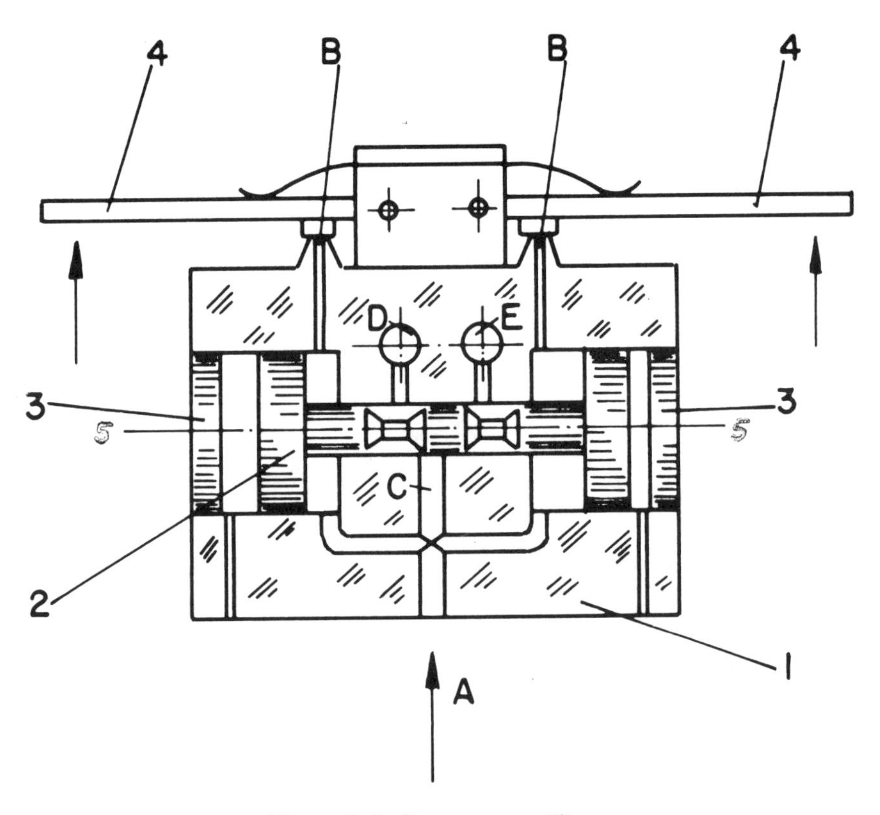

Figure 3-11 Pressure amplifier. The analyzer actuates the levers 4, thus changing the balance of the pressure in the chambers. The plunger 3 moves, and in keeping with its movement, channels D, E, C, and A are connected.

that close the openings B to the channels when there is no signal from the amplifiers.

One of the experimental devices, consisting of a receiver, analyzer, and amplifier, is shown in Figure 3-12.

In practice, for a 100-meter-long pipe, we did not observe any loss of pressure. The receivers we used are able to distinguish about 0.3 Hz. This means that in the range of 0.5 to 4 Hz, about 14 different signals can be transmitted and acted upon.

We found that there are two main obstacles that still must be overcome:

1. The system is sensitive to gas pressure in the pipes, and so some means of removing gas bubbles from the system must be incorporated.
2. A feedback device must be included to indicate the state of the valves or taps.

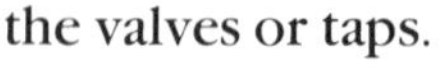

Figure 3-12 Experimental device described in Figures 3-7, 3-9, 3-10, and 3-11.

EMPATHY

This method can be considered as a particular case of the analogy method; it is an individual analogy. When employing empathy, we have to put ourselves into the situation with which we have to deal. We then have to act as though we are inside the problem and seek analogous paths of action for its solution. We can, for example, imagine that we can change our position, our size, or our proportions. Let us imagine that a colleague, John, has to solve such a simple problem as extracting a cork from a wine bottle. Put him inside the bottle, and let him take the place of the wine that "wants to get out of the bottle." After a number of manipulations, he can position himself along the length of the bottle, resting his legs against the bottom and his hands against the cork. Then by exerting himself to the limit, he can release the cork. We can thus draw the conclusion that a force from inside the bottle is able to open the bottle. What can the nature of this force be in reality? The solution is simple. It is pressure. Create pressure inside the bottle and the cork will pop out. This pressure can be generated by inserting a hollow needle through the cork and introducing air into the bottle by means of a syringe.

Empathy is a very specific method that is useful only in a limited number of cases. It is most often applied as an auxiliary method. Let us, nevertheless, look at one more example. The problem is to pass information from a rotating body to a resting one, without the use of sliding contacts. An empathy session may sound as follows:

> I am sitting on a wheel; all is going around and there is no point at rest. The only more or less stationary place is the center of the wheel. I have an insuperable desire to grip it. I could keep my hands on the shaft, but they slip around it. Perhaps I could pull a wire through a hollow shaft. But the wire will twist. This solution works only with a limited number of revolutions of the wheel, but in our case the wheel makes an indefinite number of revolutions per unit of time. It would be good to have a thin, liquidlike wire. What about a cup of mercury or some other molten metal? This could be a solution—no slipping, no friction. But there are two

> limitations: the plane of rotation has to be horizontal, and mercury is harmful to human beings. Something else? I can scream from the wheel and I can hear someone else screaming. I can blink with a flashlight or a lamp, but how can I stop the rotation of the beam? How about a mirror? I will put a mirror above the center of the rotating wheel and direct the beam along the hollow shaft. The beam does not twist, since it is a light beam. Light is an electromagnetic phenomenon as radio is. Stop! Radio—this opens the possibility of transmitting information from the rotating body and back; moreover it opens the possibility of supplying energy to equipment located on the wheel.

The answers to a lot of questions are not yet clear, but the principle on which the solution is based has been established. It is possible that this concept will be too expensive or too cumbersome, or that too much effort will have to be invested to make the concept applicable, but at least the possibility of a specific kind of solution has been discovered and analyzed.

INVERSION

Inversion is a very powerful tool in the creating of new ideas. We have already discussed it in our description of the principles used for developing new solutions. We also encountered it as a tool in the use of analogies. At this stage we will confine ourselves to a brief consideration of inversion at different levels. Here we must keep in mind the fact that this method is an effective means for the seeking of new concepts in engineering.

An example of inversion was the discovery in 1880 by Pierre and Jacques Curie of piezoelectricity—the generation of electric charges in a nonconducting crystal subjected to pressure and, conversely, the change in volume of certain crystals subjected to an electric field. One can see that it is worth keeping any thought "under suspicion" for the potential possibility of inverting it and thus obtaining a new effect, product, or solution. In the Curie case, the gap between the direct and inverse phenomenon was

covered simultaneously. In the case of Oersted and Faraday, it took about ten years for the inverse effect to be discovered.

We should not think that this creative tool is applicable only to physical laws. It also can be applied to more abstract problems, as the following example illustrates. How can the windows of a workshop be cleaned properly so as to provide proper illumination? It is apparent that the main problem is that of illumination. It is therefore worthwhile to invert the approach. Rather than spending money and effort on a complicated mechanical solution (window cleaning), it might be better to provide optimal electric lighting in the workshop. At least the pros and cons of the two approaches should be weighed.

Another example: How can a spiral electric resistance be produced? This resistance may be obtained directly by winding a wire on an insulator (see Figure 3-13a), or the inverse solution may be found by taking an insulator covered with conducting material and cutting out a spiral groove (see Figure 3-13). The inverse approach allows us to deal with very thin layers of conducting materials, and by changing the cutting pitch, high values of the resistance can be obtained. The latter process appear to be cheaper and more effective than the former one.

Further examples of inversion will be brought up during subsequent discussions.

MORPHOLOGICAL APPROACH

The morphological approach is based on the simultaneous consideration of as many situations as possible. By application of this method, the designer tries not to miss any opportunity in the "field of solutions." The morphological approach fits both the "what" and "how" problems. Let us consider an example of its application to the "what" problem, in this case, one concerning the domain of car users. The simplest case can be illustrated with the help of Table 3-1.

Some factors relevant to car users appear in the left-hand column and again at the top of the table. We now have a two-dimensional domain of possibilities. (*Note:* The domain is symmetric relative to its diagonal.) In analyzing combinations of the factors, we may obtain, for instance, the following:

1. Fuel-consumption meter (dashboard + fuel consumption).

2. Tire-pressure meter (dashboard + tire pressure).

3. Exhaust-gas-composition meter (dashboard + exhaust gases).

4. Device for inflating tires with exhaust gases (exhaust gases + tire pressure).

5. Device for cleaning by the use of exhaust gases (exhaust gases + cleaning).

6. Inflatable covering blown up by the exhaust gases (exhaust gases + covering).

7. Indicator of the state of the ignition system (dashboard + ignition).

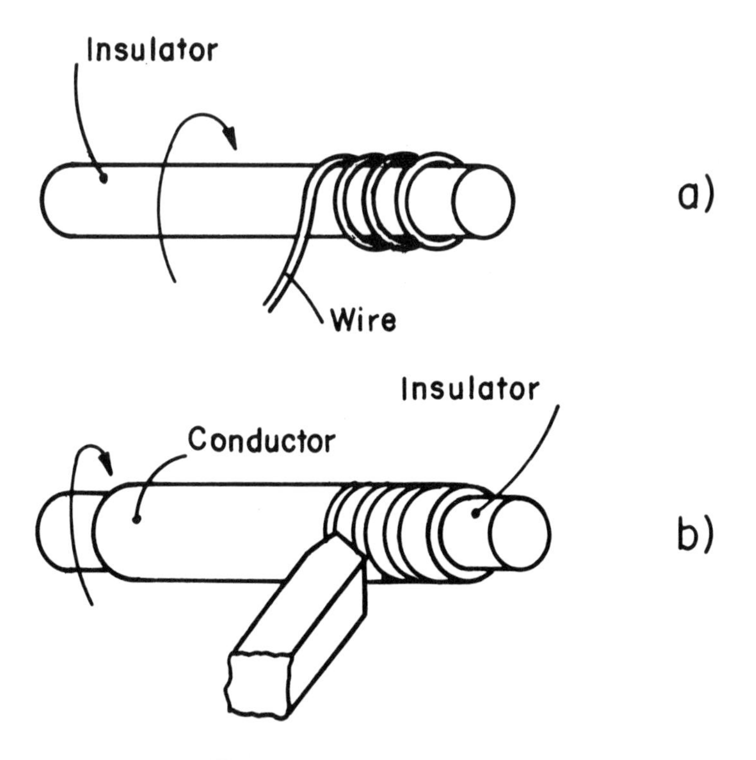

Figure 3-13 Two ways (direct and inverse) of producing electrical resistance. In (a) a wire is wound on an insulator and in (b) a spiral groove is cut into a conducting material (say carbon), which has been laid over an insulator.

Table 3-1
Morphological Approach

8	*1. Speed*	*2. Distance*	*3. Fuel consumption*	*4. Tire pressure*	*5. Exhaust gases*	*6. Cleaning*	*7. Cooling*	*8. Dashboard*	*9. Ignition*	*10. Covering*
1. Speed										
2. Distance										
3. Fuel consumption										
4. Tire pressure										
5. Exhaust gases										
6. Cleaning										
7. Cooling										
8. Dashboard										
9. Ignition										
10. Covering										

At this stage we do not need to decide how to realize these proposals, but rather whether there will be any demand for them and thus whether it is worthwhile to develop them. Some of the potential products are discussed later.

Let us now pass on to the "how" stage. For this purpose the designer must first ennumerate all the possible directions available, and then must consider all the steps in each direction and all the elements of each step. The combinations of elements then must be analyzed. It is easier to understand the application of the morphological approach to the "how" problem in terms of an example. For this purpose we will consider a fuel-consumption meter. The function of this device is, by definition, the continuous supplying of data to the driver about the fuel consumption of the car per unit of distance—liters per kilometer or gallons per mile. Of course, the inverse values—kilometers per liter or miles per gallon—are also acceptable. It is clear from the definition of the device that its design must include an aspect that treats information about the amount of fuel burned and an aspect that deals with the distance traveled by the car. In this case we met with difficulties in solving the first aspect, which is why we are using it in our discussion.

In our example the directions for solution seeking are as follows:

1. The structure of the flowmeter.
2. The way in which the measured results are to be indicated.
3. The means of computation.

To simplify the consideration, we stopped at three variables, that is, at the three-dimensional representation. With regard to the structure of the flowmeter, we can imagine a purely mechanical approach, an electric one, a hydraulic one, and so on (Figure 3-14). The kind of display can be chosen in accordance with Figure 3-14—digital, analog, acoustic, or colorific. For the computation method, continuous, interrupted, and analog approaches can be considered. The three-dimensional volume of the solution space is covered by 54 (in this case) possible combinations, from which, after thorough analysis, the most promising one can be chosen.

The solution space shown in Figure 3-14 can be represented in a different manner, as shown in Table 3-2. Here the multidimensional "spaces" can be better and more simply portrayed.

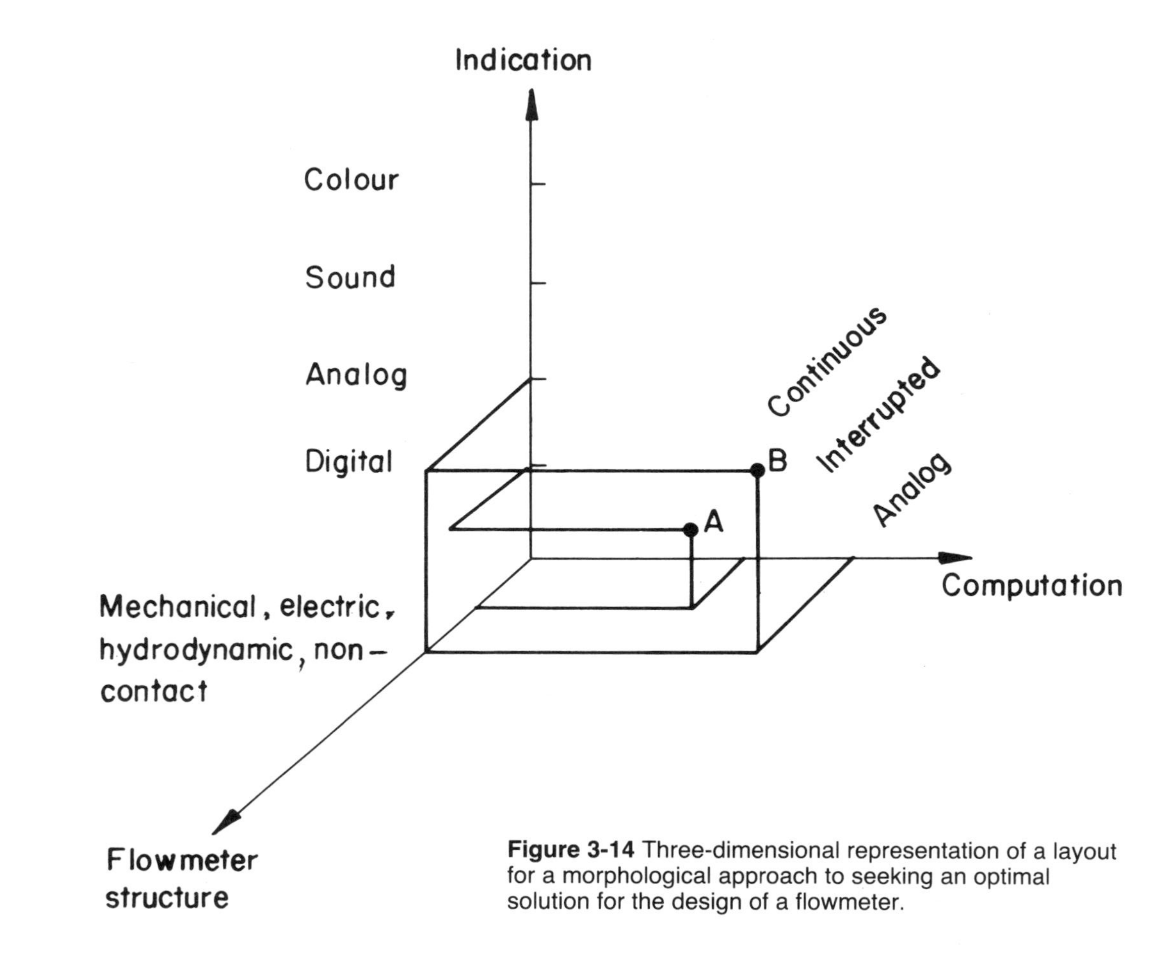

Figure 3-14 Three-dimensional representation of a layout for a morphological approach to seeking an optimal solution for the design of a flowmeter.

Table 3-2
Example of "Solution Space"

Direction	*Step I*	*Step II*	*Step III*	*Step IV*
		IV		
1. Flowmeter structure	Mechanical	Electric	Hydrodynamic	Noncontact
2. Means of indication	Digital	Analog	Acoustic	Calorific, etc.
3. Computation	Continuous	Interrupted	Analog	

Note the following points:

1. The morphological method does not add information to the problem, but it puts the considerations in better order.

2. If 1, . . . , i, . . . , k is the number of consideration directions, and A_i is the number of steps on the *i direction, then the number of possible combinations will reach*

$$N = \prod_{i=1}^{k} A_i$$

3. If N is large, the method ceases to work.

SYSTEMATIC APPROACH

The systematic approach is based on a previously mentioned idea that it is better and more effective to solve a problem by moving from the end to the beginning than vice versa.

If the final result is properly formulated, then the problem can be solved in the best possible way. Serious consideration, therefore, must be devoted to this first and very important stage of problem solving. An extremely simplified idealized form of the final aim has to be formulated and imagined. This step is not as easy as it seems, some psychological obstacles first must be overcome.

Rules for Formulating the Problem

1. The first rule we have to remember is that the formulated problem must not sound like

To change A and B for achieving C and D

or like

To change A and B

but rather like

To achieve C and D

Let us consider the following example. The higher the speed of a car, the smaller will be the traction between the wheels and the road, which is an undesirable factor. In the context of the example, let us compare the following formulations of the task.

a. To increase the weight of the car to increase the traction of the wheels.

b. To increase the weight of the car.

c. To increase the traction.

The first two statements are wrong because they automatically exclude, for example, the aerodynamic possibility (Figure 1-1). These two statements would be better phrased as:

a. To increase the pressure of the wheels on the road to increase the traction.

b. To increase the pressure of the wheels on the road.

But the best formulation remains the third (i.e., the most general) statement. This formulation does not exclude the possibility of putting chains on the tires, of using sand, of designing special-purpose tires, of adding wheels, and so on.

2. The second rule may be briefly stated as: When formulating the aim *do not* think how or by what means you will reach it.

3. The third rule may also be briefly presented: Do not *guess* whether or not it is possible to solve the problem and thus attain the aim.

4. The fourth rule requires a short explanation: Do *not* be afraid of solving "old" or "perpetual" problems. Often the older the problem, the easier it is to solve. The explanation for this is simple: The complexity of the problem does not change but the means for solving it (knowledge, technology, tools, etc.) *do* develop. This rule perhaps can be best expressed in the words of Jules Verne (1828—1905), who wrote that all that one human being can imagine, others will succeed in carrying out in reality. We should remember that many of Jules Verne's fantastic predictions have been realized, or even surpassed.

Definition of the Approach

The systematic approach offers a certain system for the seeking of solutions or a definite sequence that should be followed for overcoming contradictions. In simplified form the process may be described as follows:

1. Define the ideal final aim or result. What do you want to obtain in the optimal case?

2. Define the factors that may complicate the achieving of the ideal result. What is (are) the obstacle(s) on the way to success?

3. Define why the obstacles present a problem. What is the direct cause of the stumbling block(s)?

4. Define the conditions under which nothing will prevent the achievement of the desired result. What conditions will cause the obstacles to disappear?

Do not think that by following this sequence you will automatically succeed. Obviously reality is more complicated. For instance, the first step in this list requires a certain thinking procedure, which itself consists of some auxiliary steps.

Auxiliary Questions in Problem Formulation

1. What is the final aim of the sought solution?
2. Is it possible to reach the aim in a roundabout manner?
3. Which way appears more promising—the direct or the roundabout?
4. What are the major and minor requirements that the solution must provide?

The implications inherent in these questions can be understood in light of a detailed example: The problem we will consider is how to indicate to a driver whether or not the pressure in the tires of the car is correct. This would facilitate the prevention of the danger caused by a sudden flat tire. Let us follow the foregoing sequence.

1. Final aims:

a. To indicate the state of the tires, that is, to indicate the value of the pressure in them.

b. To indicate whether the pressure is less than a certain allowed value.

2. Definition of the roundabout solution:

a. To indicate the distance between the drum of the wheel

and the surface of the road, which is obviously correlated with the pressure.

b. To indicate the deformation of the tire.

3. Direct or roundabout way: At this stage the roundabout way does not appear attractive—potential roundabout solutions do not seem better than a direct solution.

4. Requirements:

a. The price of the device built in accordance with the solution must be reasonable.

b. The device must be applicable to existing cars.

c. The indicator showing which tire is bad must be both acoustic and visual.

d. A minor requirement: An accoustic indication of a bad tire is sufficient; the driver is responsible for defining which is bad by climbing out of the car and seeing which tire is flat.

Now we can pass on to the solution-seeking procedure—steps 1 through 4. This procedure may be repeated several times, each time narrowing the domain of the search. Thus the "first round" will take the following form.

1. The ideal final result: An accoustic and visual indicator of "dangerous pressure in the tires" would be installed on the dashboard.

2. Complicating factors: A mechanical solution with the installation of a complicated piping system connected to the tires by means of rotating sealed couplings on the wheels is not acceptable because this solution requires that the wheels have a special design.

3. Cause of the obstacles: The mechanical solution is too expensive and is not applicable to existing cars because of the need to interfere with the mechanical structure of the wheels.

4. Solution to the obstacles: The mechanical concept must be changed for, say, an electrical one.

The "second round" will narrow the problem as follows:

1. The ideal final result: An electronic "dangerous-pressure-in-the-tires" indicator would be installed on the dashboard.

2. Complicating factors: An electric pressure transducer in each tire creates an electric signal that has to be transferred to the dashboard. The difficulties are the powering of the electronic circuit on the wheel and the receiving of the information from the wheel.

3. Cause of the obstacles: The difficulties arise from the rotation of the wheel. To overcome this obstacle, we need to find a noncontact and nonwired electric connection. (See the example under "Empathy.")

4. Solution to the obstacle: A radio-wave connection can be used for both energy feeding and information receiving purposes.

The proposed solution is shown schematically in Figure 3-15. A pressure relay PR is installed in the valve of the tire. When the

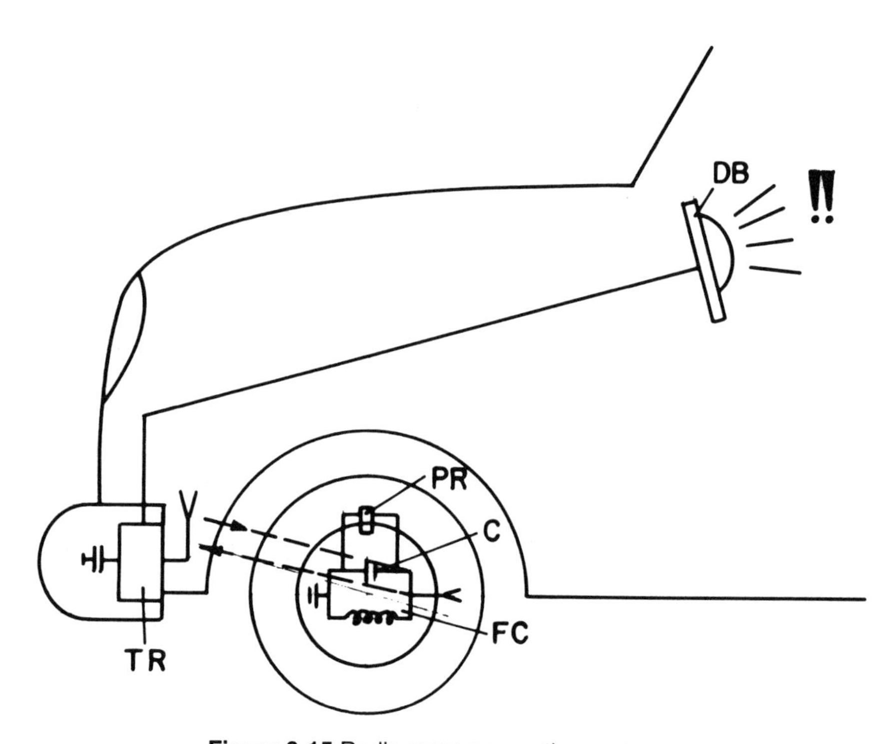

Figure 3-15 Radio-wave connection system for a "dangerous-pressure-in-the-tires" indicator located between the wheel and the dashboard.

pressure in the tire is correct, the contacts in PR are closed and thus short-circuit the capacitance C. When the pressure decreases and reaches dangerous values, the contacts of the PR open and the circuit of the wheel becomes a frequency loop with natural frequency ω. A transmitter TR fastened to the bumper continously transmits radio waves of frequency ω to the frequency loop FC. Thus when the pressure drops, the power used by the transmitter changes as a result of resonance, and this event appears on the dashboard DB in the form of sound signals and (if desired) as a visual indication of which tire is flat.

AFTER THE SOLUTION IS FOUND

Almost all methods of generating solutions are suitable for solving both the "what" and the "how" problem. This is illustrated in Table 4-1.

These methods are suitable for the initial stages of creation, but there is another side to the question. This aspect, which we call synthesis, becomes apparent only after a solution has been found, or even after a device has been built. Synthesis is made up of three components—the "expansion" principle, the "complementary" principle, and the "excess" principle.

Table 4-1
Comparison of Solution Methods

Method	*What*	*How*
Trial and error	+	+
Brainstorming	+	+
Analogy	+	+
Empathy	–	+
Inversion	+	+
Morphological approach	+	+
Systematic approach	–	+

THE EXPANSION PRINCIPLE

To clarify the expansion principle, let us reconsider the example of a fuel-consumption meter for cars. The device (which measures two values in a digital manner and then divides one value by the other) comprises a flowmeter for relatively low flow rates (0.15–0.25 cm^3/sec) and a distance meter. Let us suppose that we own such a device and ask ourselves what new ideas we can extract from it and how we can propagate the ideas and solutions peculiar to it to other technical fields. We call this approach the *expansion principle.*

Such a flowmeter, which is relatively cheap and sensitive, can be used in any situation in which digital information about low flow rates of any liquid is required. Gas flows can also be measured. In our case the same device, with minor modifications, can measure air consumption at rates around 2.5 cm^3/sec. Thus what we have really invented is a universal flowmeter, and our next step should be a comparison with existing equipment of this type and examination of the demand and marketing conditions for these devices.

We also have designed an instrument in which the distance covered is divided by the fuel consumption, or vice versa. Why not seek other applications in which the consumption of a fluid has to be divided by any other physical value—time, speed, or even the consumption of some other substance?

Let us take another example, the tire pressure indicator. That solution was based on the "separation" of the rotating part of the car from the source of energy supply and information gathering. Where else could this type of radio solution be useful? The answer, of course, is for any other rotating object. Besides this particular case, it could also be applied to any process taking place in a closed space in which there are limitations on direct means of communication, for instance, in the burglar alarm system for automobiles presented in Figure 4-1. When someone tries to break into the car, the mechanical oscillations cause the alarm relay AR to short-circuit the LC circuit, and this action changes the energy balance of the transmitter. The latter component then actuates an alarm device located, for instance, in the car owner's home.

As another illustration we can consider the expansion of "Ru-

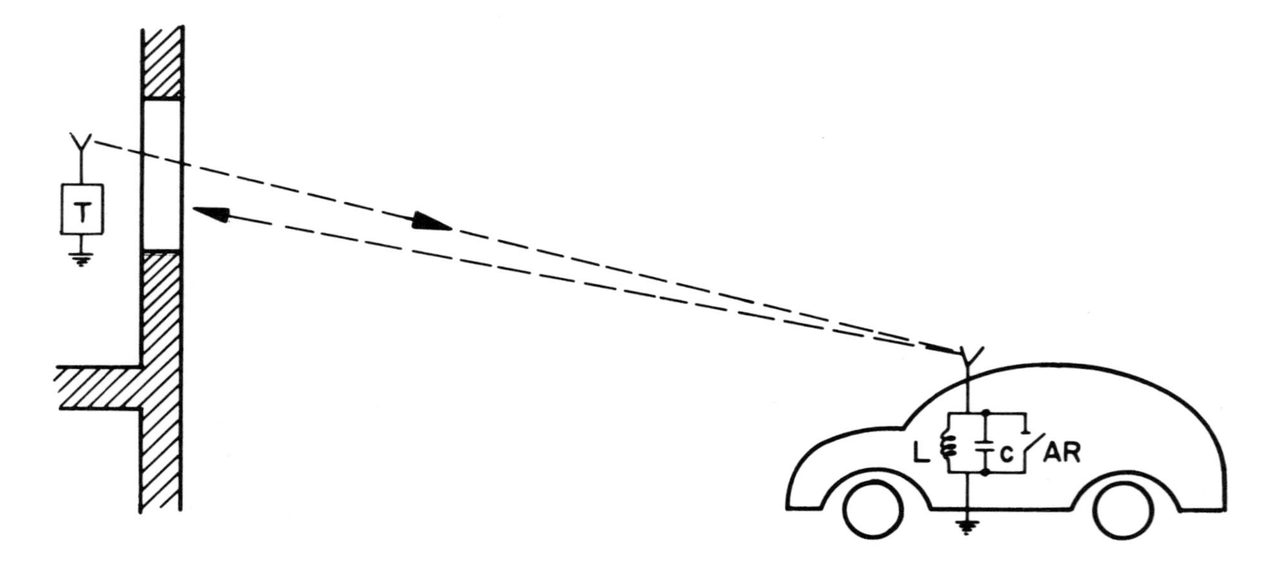

Figure 4-1 Burglar alarm system for automobiles. The alarm device in the car owner's home is connected by radio to a transmitter in the car.

bik's cube." One direction of expansion may be to change the number of constituent small cubes (which we call cubicles). Another direction may be to alter the shape. Why only a cube? What about a pyramid or a sphere? (Psychological inertia). For instance, if we take a 4 × 4 cube and round its apexes, we produce a sphere. This operation gives the game new properties. Let us print a map of the earth on this sphere. By changing the relative positions of the eighths, we obtain a perfect illustration of our topsy-turvy world and its political ambitions (Figure 4-2). Furthermore, we can print any portrait on the surface of the sphere, thus transforming the dry mathematical exercise of the initial cube into an amusing game that can fascinate many different types of people.

We can expand the concept even further by asking: Why not use a definite form? Figure 4-3 shows some of our fantasies as to how to execute this idea. Imagine that the small attractive elephant or pig in the picture could each appear in approximately $2.6.10^8$ combinations. Is this not fascinating? Try to move the parts and then put them back in order!

Figure 4-2 A topsy-turvy world built on the surface of Rubik's cube. A 4 × 4 cube with round apexes forms the basis for this puzzle.

Another possibility of applying the expansion principle is to combine (Principle III) the "cube" with another game, for instance, "15" (the Samuel Lewis game). Let us propose a cube with 2×2 cubicles on each facet, that is, eight cubicles altogether, which creates $8! \times 3^8 = 2.6453 \times 10^8$ combinations. The six visible facets of the cubicles consist of $6 \times 4 = 24$ squares on which we place, by special constructive means, 23 movable numbered squares. By moving the numbered squares through the empty square and turning the cubicles, we create a new entertainment by bringing the numbers into the desired order. This yields another $23! = 2.6.10^{22}$ combinations.

THE COMPLEMENTARY PRINCIPLE

The analysis of a new product often leads to the revealing of some previously unknown properties of the product. One must think these new facts over and then decide whether they can be useful. When the characteristics of the flowmeter described earlier were reexamined, it became evident that there was a definite relationship between the viscosity of the liquid flowing through the device and the slope of the line in a rotation versus flow-rate plot. The thought occurred to us the device could be used as a viscometer. Thus we can see that the utilization of the newly discovered properties of a device forms the basis of the comple-

Figure 4-3 Why should Rubik's cube have a cubic shape? These figures could each appear in approximately $2 \cdot 6 \cdot 10^8$ combinations.

mentary principle; that is, some complementary properties of a solution are applied to solving another problem. For instance, the same fuel-consumption meter can also be used to remind the driver that the time has come to change the lubricating oil. The information on the distance traveled is, in any case, stored in the memory of the device, so there is no difficulty in arranging for a visual signal to appear on the dashboard every 3000 km (or any other distance)—"change the lubricant!" This is a complementary aspect of the solution to the main problem.

Let us look at another example of the complementary principle, this time the design of an accelerometer. When engaged in an experiment with reproducing analog images by means of carbon paper, the author noticed that the contact pressure influenced the results. This property (which is easily explainable) led to the creation of an inexpensive accelerometer (U.S. Patent no. 4114453).

The structure of this accelerometer is shown in Figure 4-4A. It consists of a housing 1 in which an inertial mass 2 is placed, variable resistances 3, a casing 4, and electric wires 5. In this example a simple bridge layout is used for the measurement of the acceleration (or force). The electric voltage corresponds to the acceleration components A along the coordinate axis X. A battery B provides the voltage for the bridge. Changes in the voltage of the battery will influence the measuring scale and the sensitivity of the device. A possible solution for the construction of the variable resistances is shown in Figure 4-4b. It can be seen that basically the variable resistance consists of a thin insulator film 7 (for example, paper) covered with a conductive or semi-conductive layer 8 (for example, carbon). In this case the film is criped so that one end of layer 8 comes into contact with the inertial mass 2 and the other with the cover 4.

THE EXCESS PRINCIPLE

To help us understand this principle, let us look at Figure 4-5, which shows schematically the structure of an air compressor that is actuated by the exhaust gases of a car (see Problems 4 and 5 in the discussion of the morphological approach). This compressor

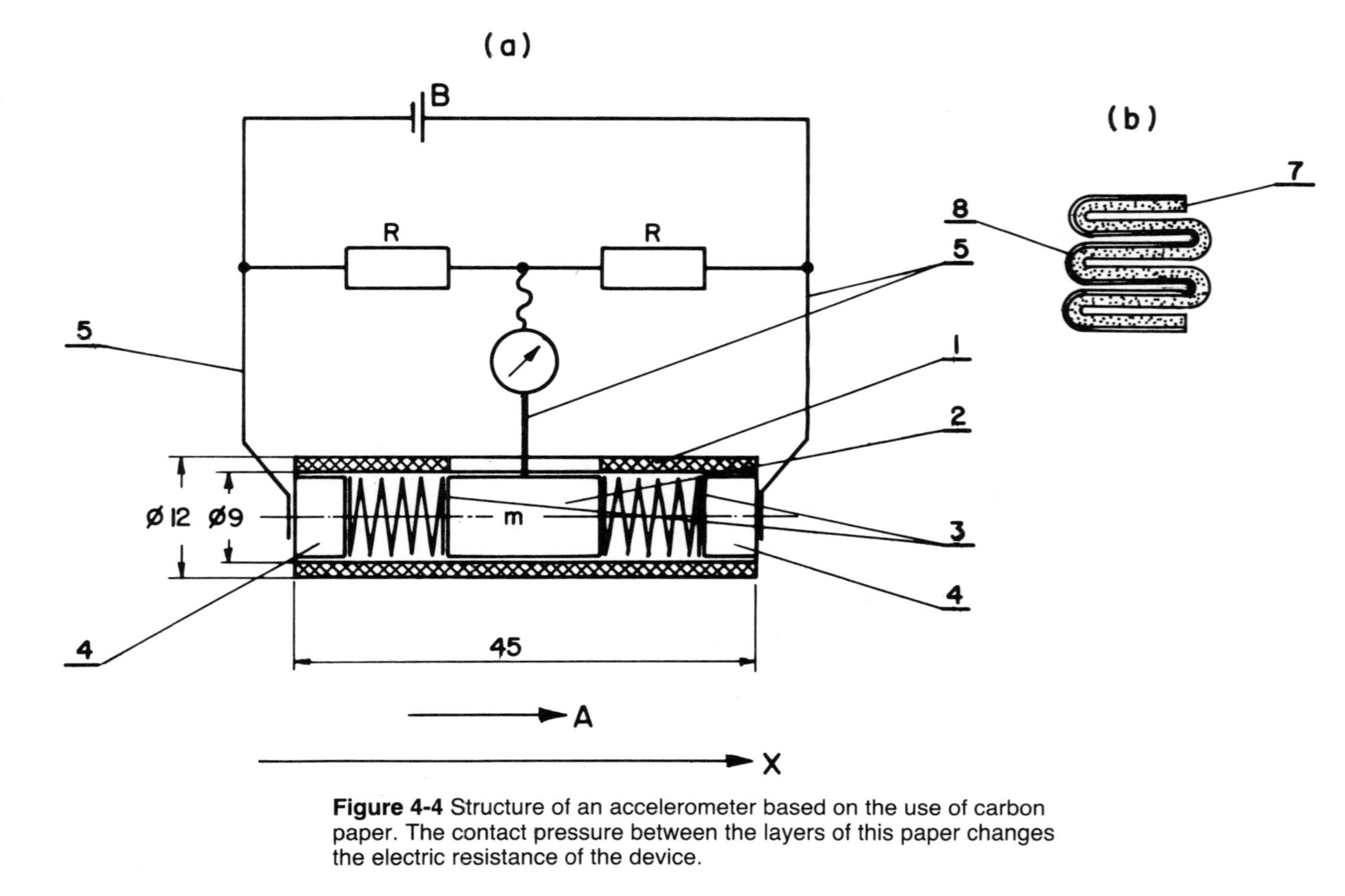

Figure 4-4 Structure of an accelerometer based on the use of carbon paper. The contact pressure between the layers of this paper changes the electric resistance of the device.

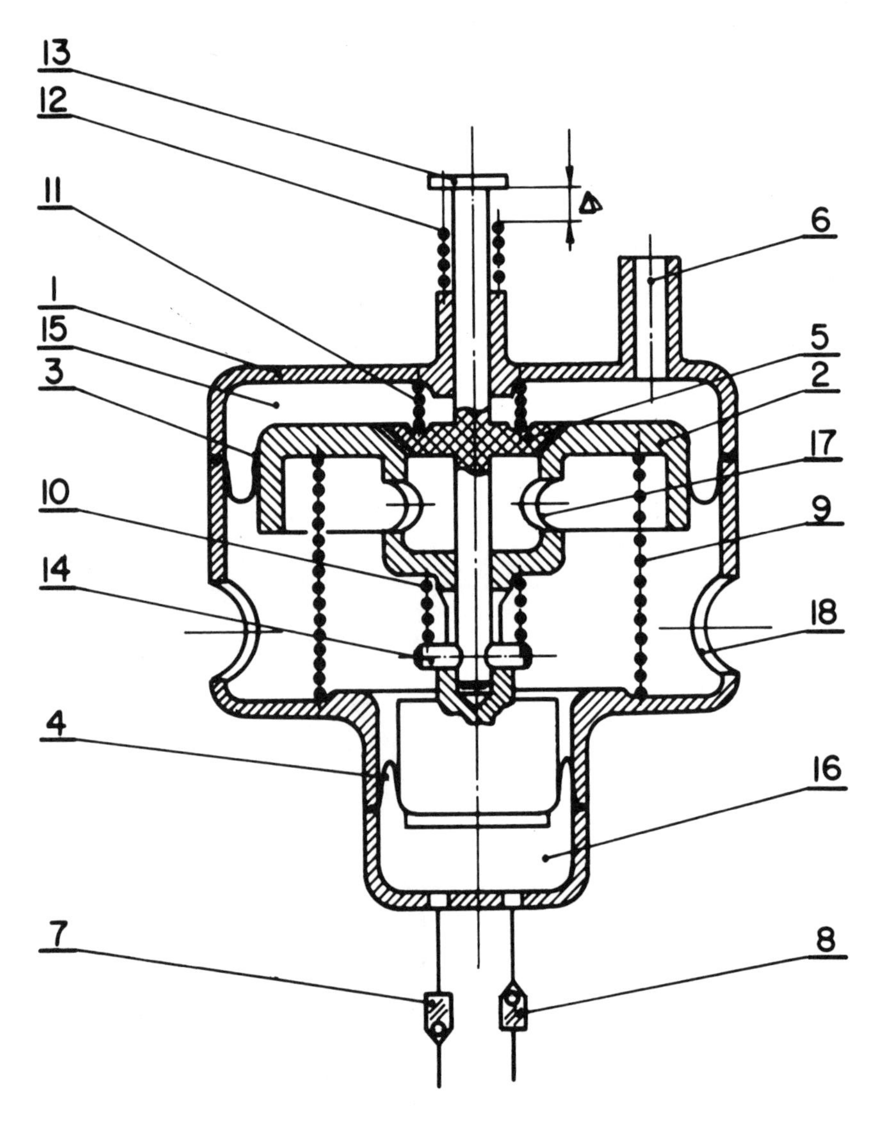

Figure 4-5 Structure of an air compressor actuated by the exhaust of a car. This device can be used for inflating tires as well as for actuating other pneumatic equipment.

utilizes car exhaust gases for the production of compressed air. The figure shows that the compressor consists of the following components: 1—housing; 2—plunger; 3—"large" diaphragm; 4—"small" diaphragm; 5—valve; 6—low-pressure inlet; 7—atmospheric air inlet valve; 8—high-pressure outlet valve; 9, 10, 11, and 12—springs; 13 and 14—supports; 15—low-pressure chamber; 16—high-pressure chamber; 17 and 18—openings.

When the low pressure is connected to the inlet 6 and chamber 15, the plunger 2 begins moving downward as a result of the pressure of diaphragm 3 on the spring 9. At this time the valve 5 is closed because of the pressure in the chamber 15, the action of spring 10 on support 14, and, at the very beginning of this stroke, the action of spring 11. During the downward stroke, the small diaphragm 4 compresses the air in the chamber 16 and forces it through the outlet valve 8. When the plunger 2 has moved a certain distance, the support 13 reaches the spring 12. When its deformation becomes considerable, the spring 12 opens the valve 5. At this moment the low pressure rapidly disappears because of the low aerodynamic resistance of the openings 17 and 18. This results in the lifting of the plunger 2 due to the action of spring 9. The small diaphragm 4 sucks fresh air into the chamber 16 through the inlet valve 7. The correct choice of the mass of the plunger 2, the stiffness of the spring 9, and the geometry of the openings will facilitate a close-to-resonance working regime of the device. The outlet pressure and the productivity of the device will depend on the ratio of the arcaas of the diaphragms, on the input pressure, on the consumption of the exhaust gases, on the volumes of chambers 15 and 16, and on the dynamic properties of the moving system. In our case the input pressure is about 0.4–0.6 psi and the output pressure is about 28–32 psi, the productivity being about 3.5 liters per minute. By changing the parameters of the device, a different set of figures can be obtained.

At this stage a word of warning is in order. When a new idea appears, do not become too optimistic; first check whether the same new effect cannot be achieved in a simpler way. It is, however, a general rule to examine whether a new solution may contain more possibilities than are required. For example, when we apply the expansion principle to the air compressor, we must conclude that we also have a vacuum source (via the air inlet

valve). The next step naturally would be the creation of a vacuum cleaner actuated by the exhaust gases of a car. Here we must be careful! We must examine other possibilities, and perhaps we will find something cheaper and easier to produce. The type of "danger" inherent in the application of the excess principle has its roots in psychological inertia. Figure 4-6 shows a simpler solution based on the ejector principle. In this design there are neither moving parts nor superfluous properties.

To deepen our understanding of the excess principle, let us look at another example—a "swimming" device. There are many different kinds of swimming devices driven by human power, most of which are propelled by paddle wheels. This means that a person must drive a mechanical transmission cumbersome enough to maintain the rotation of the paddle wheels, the latter driving the water backward while the craft travels forward.

Why do we need an intermediate agent—the transmission? All we need to provide is a stream of water, or a water jet, in the backward direction. Thus the idea of a swimming craft propelled by a human-driven water jet was born. Calculations show that the speed of such craft is about 1–1.5 m/sec, taking into account the fact that the capacity of an average person is about 0.2 hp (or even 0.4 hp for less than 2 minutes of action). Figure 4-7 shows a possible design for such a swimming device, which consists of a plastic or rubber float 1 and a propulsion unit 2. Figure 4-8 shows the structure of the propulsion unit comprising two bellows

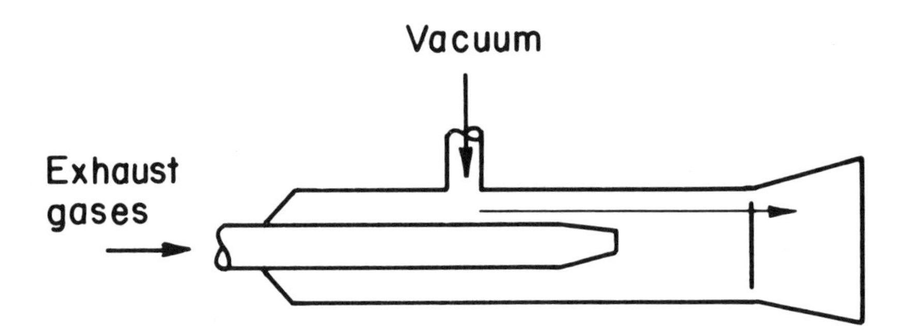

Figure 4-6 Vacuum generator actuated by the exhaust of a car is based on the ejector principle.

1, which can be activated independently or together. In both cases valves 2 and 3 ensure that the propelling water flow is directed rearward.

The traveling speed can be increased by enlarging the driving power. In accordance with Principle VII of creativity, let us put a number of people on the float, thus increasing the power. Note that if the cross section of the flat does not change (as is the case when the people sit one behind the other), the speed will be approximately proportional to $\sqrt{n}$, where n is the number of people. There is, however, the possibility of increasing the speed in another way, by copying the idea of the bow and arrow (Principle I), by accumulating the energy. The layout of such an energy-accumulating propelling device is shown in Figure 4-9. The bellows 1 pumps the water via valves 2 and 3 into an elastic bag, thus compressing the air in a solid vessel 5. Then, at a given moment, the driver frees the water via valve 6 and it flows through the nozzle 7.

Further increases in speed demand other energy sources, and this is where the excess principle can be applied. Vessels propelled by water jets currently are built according to the layout shown in Figure 4-10. The water is sucked in through inlet 1 by means of a compressor or pump 2 and is ejected from the chamber 3 through the outlet 4. The pump is driven by an engine 5 (for example, an internal-combustion engine). By applying the excess principle to this design, we obtain a solution based on the shortening of the process from the combustion of the fuel to the expulsion of the water from the jet. In this solution we are aiming

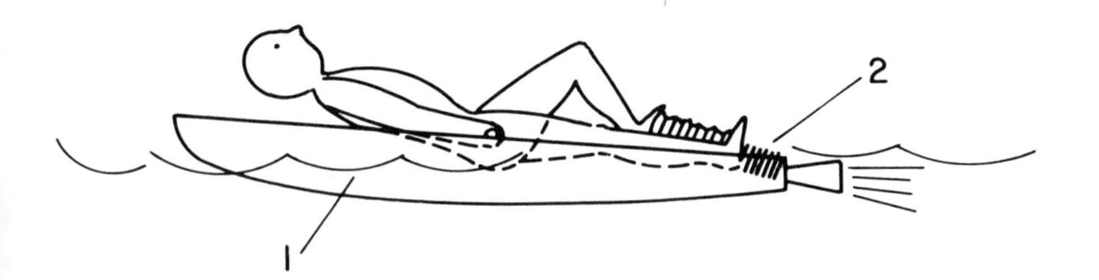

Figure 4-7 Manually actuated swimming device propelled by water jets.

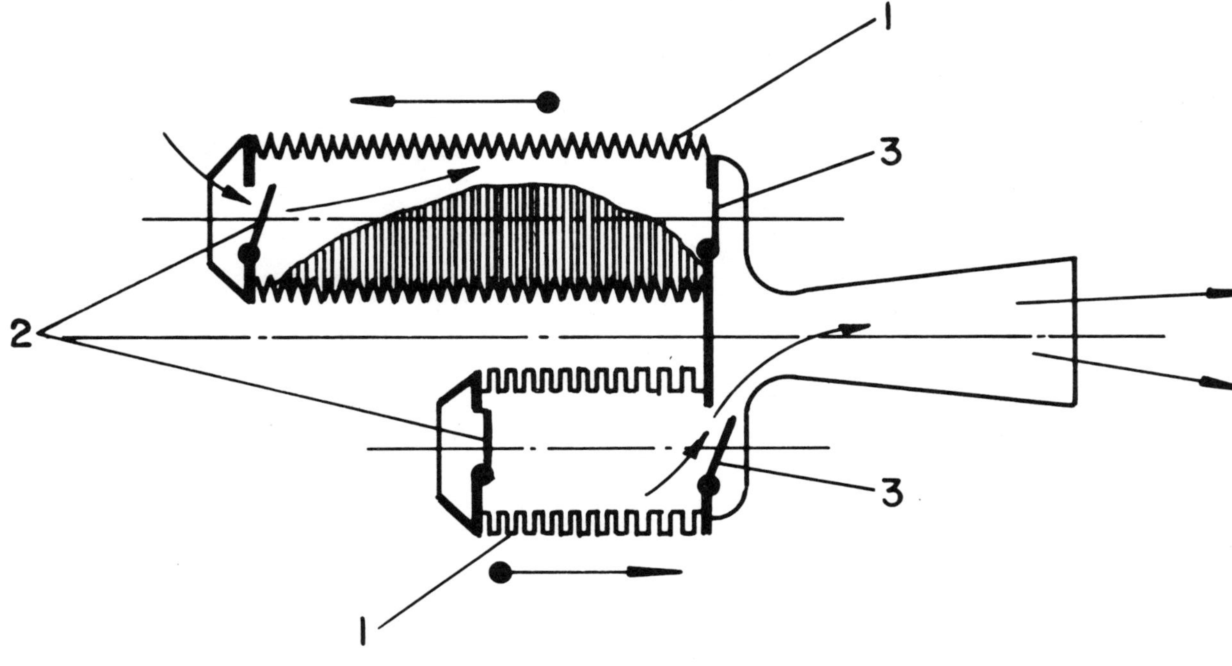

Figure 4-8 Structure of the propulsion unit of a swimming device in which the two bellows can be activated independently or in concert.

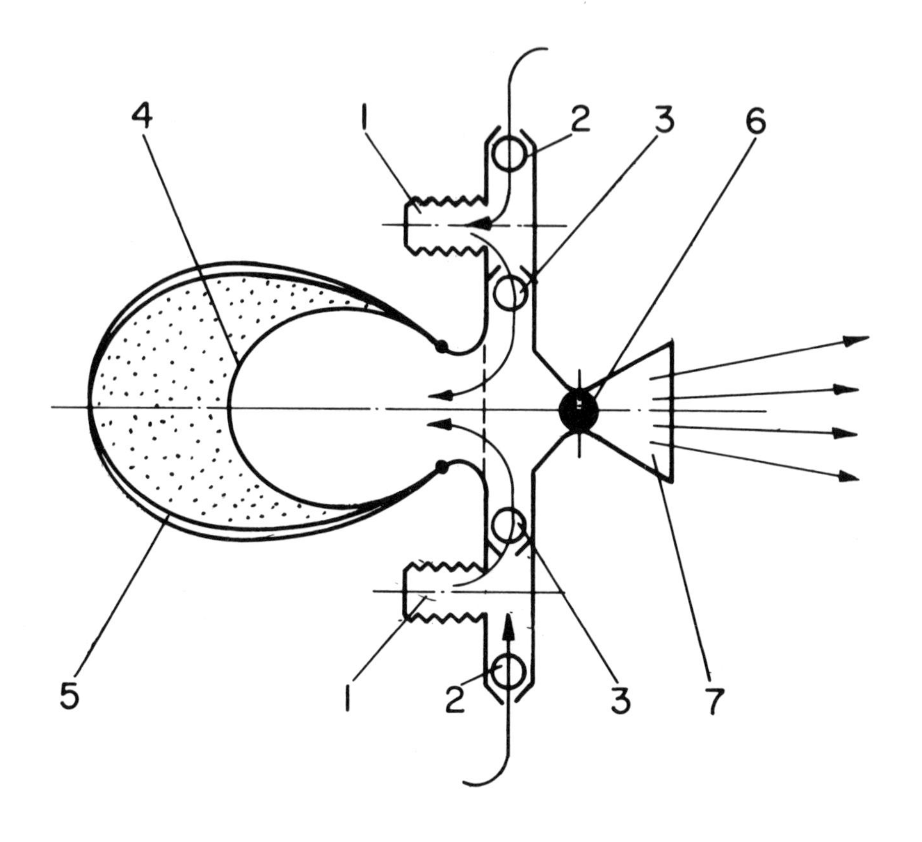

Figure 4-9 A propelling water jet provided with a pneumatic energy accumulator. At a given moment, the driver frees the water via valve 6, increasing the traction of the swimming device.

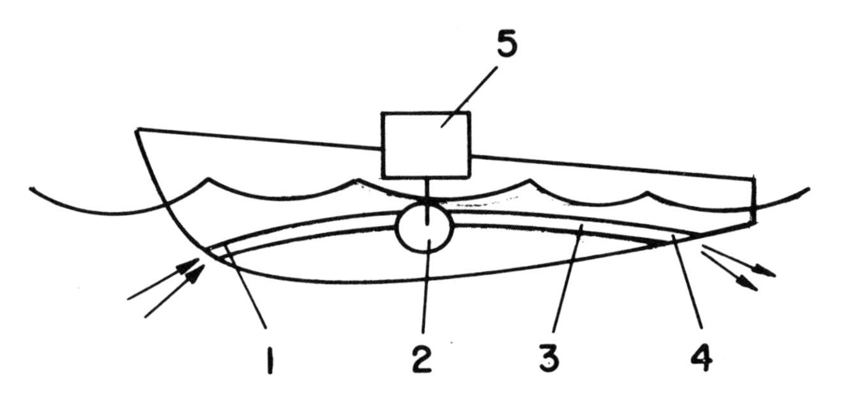

Figure 4-10 Layout of vessel propelled by water jets.

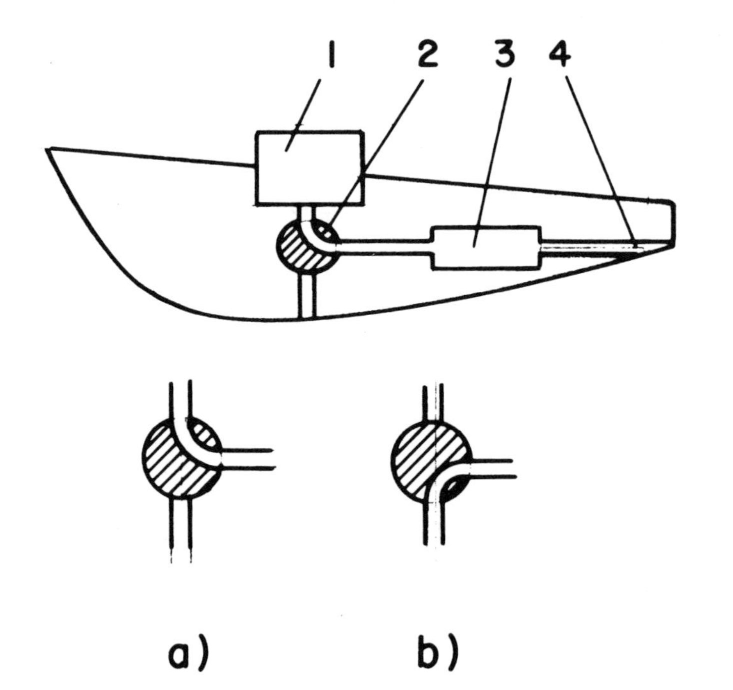

Figure 4-11 Layout of a vessel driven by a water jet where the propulsion is obtained by the direct action of the expanding gas (generated by burning fuel) on the water.

a. The valve in the propulsion position.

b. The valve in the gas-generating position.

at finding a means of expelling a particular volume of water per unit of time. To this end we propose to burn a fuel to produce a gas that will, in a direct manner, push the water out of the jet chamber 4. This concept is presented schematically in Figure 4-11a. Gas source 1 is connected to the jet 4 via a valve 2 (which has two positions, *a* and *b*). The chamber 3 is filled with water through valve 2 in position *b*, and when the pressure reaches a certain value in the gas source 1, the valve 2 will move to position *a*, and the water will be pushed out, causing traction. When the chamber 3 becomes empty, the valve 2 moves into the *b* position, providing the time for the chamber to fill with water and for another batch of gas to be produced from the burning fuel in the gas source 1.

CONCLUSIONS

We have considered some topics relating to the processes of generating new technical (and, nontechnical) ideas and of seeking new technical solutions. We have tried to show the development of technical thinking and ways of overcoming pitfalls. Let us make it clear, however, that a knowledge of such techniques does not automatically convert the reader into a technical genius. But this way of thinking does open up certain possibilities:

1. It allows us to analyze existing technical solutions, their history, and the solutions of our colleagues and co-workers, and in this way to accumulate skills in effective thinking.
2. It allows us to organize our approach to overcoming technical problems and, at certain stages, to perfect our craft and to improve our techniques (just as a chess player improves by practicing).
3. It allows us to accumulate a number of example solutions, which (when a considerable number have been stored in our memories) will serve us in our professional lives, either directly as ready solutions or as analogies and stimulants for finding new solutions.

We recommend that you start a notebook and keep it with you at all times so that you can make brief notes of your thoughts (even seemingly stupid ideas); that you read; that you listen; and that you use sketches and layouts. Do not be embarrassed if your solution to a problem seems trivial—nobody reads your notes. When the number of notes you have made reaches 300, you will find that you are much more able to reach technical solutions. When the number approaches 500 to 600, you will be a person of technical erudition.

Although our discussion has not touched on engineering calculations, we must remember that after the solution has been conceived, we must check whether it is practical, whether it can be realized, whether it contradicts the laws of nature, and what dimensions and expenses are involved. Sometimes the concept of a technical solution occurs as a result of an analytic investigation; that is, the engineering calculations precede the creative process. For example, a two-mass system may be described by the following system of differential equations.

$$\begin{aligned} m_1\ddot{x}_1 + b(\dot{x}_1 - \dot{x}_2) + c(x_1 - x_2) + c_0x_1 &= c_0X \\ m_2\ddot{x}_2 - b(\dot{x}_1 - \dot{x}_2) - c(x_1 - x_2) &= 0 \end{aligned} \quad (1)$$

The symbols are in accordance with the schematic model given in Figure 4-12a and their physical meanings are as follows:

m_1, m_2 oscillating masses
c, c_0 stiffnesses of springs
b damping coefficient, or a coefficient that describes the rate of energy dissipation in the system
X the excitement, which is some external movement
x_1, x_2 displacements at the masses m_1 and m_2 respectively

Let us substitute $x = x_1 - x_2$ into equation (1). Then

$$\begin{aligned} &m_1\ddot{x} + b\dot{x} + cx + c_0x_2 + m_1\ddot{x}_2 = c_0x \\ &m_2\ddot{x}_2 - b\dot{x} - cs = 0 \end{aligned} \quad (2)$$

Now the dissipated energy E can be found by using the expression

$$E = b\int_0^t \dot{x}^2 dt \quad (3)$$

It is easy to imagine, as an application of such a mathematical model, a body m_1 floating on the open seas, on which another mass m_2 is suspended by means of a springlike device that has a stiffness c (Figure 4-12b). The energy is dissipated via a system that transforms the relative moment $x = x_1 - x_2$ of the masses into some kind of energy (compressed air, water, or electricity). Here the excitation X is performed by the sea waves, and the water plays the role of the second spring characterized by stiffness c_0 (in the mathematical model). Thus at least the concept of a utilizer of the energy generated by sea waves has been created. In this case the "mathematics" is the forerunner of the creation.

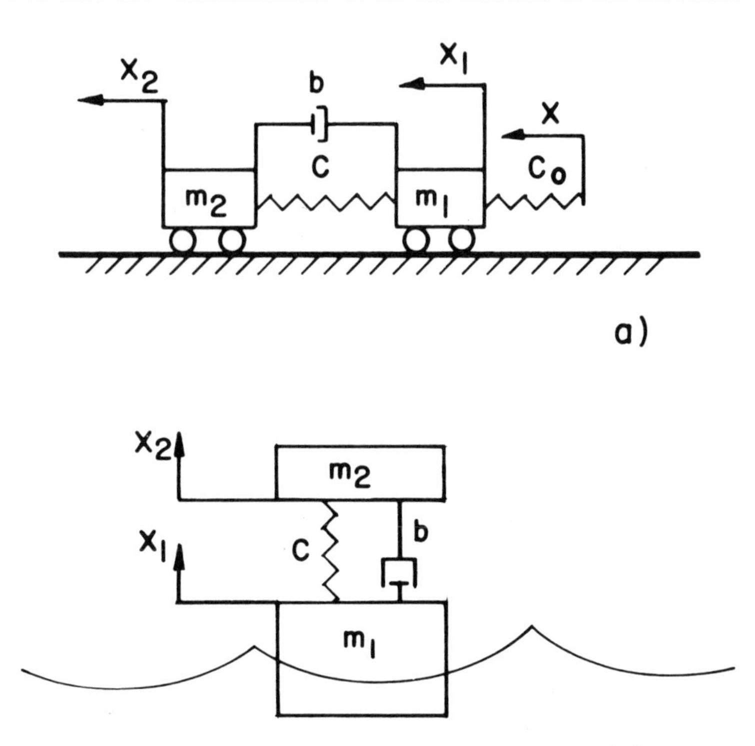

Figure 4-12
a. Model of a two-mass oscillator.
b. Application of the model to the idea of a sea-wave-energy utilizer. The relative motion of the bodies X_1–X_2 can be transformed into energy.

The invention of radio communication is another example of mathematics giving rise to creativity. The theoretical existence of electromagnetic waves—later known as radio waves—was first proposed by Maxwell in 1873. His computations, in turn, led to the experimental discovery of radio waves by Hertz in 1885, and to the first practical application of radio communication by Marconi in 1897. Similarly the diesel internal-combustion engine was invented analytically through theoretical calculations of the thermodynamic cycle.

EXERCISES

The following exercises may serve as the first steps on the road to technical creativity. Good luck!

I. Study the history of the development of creativity. Indicate which principles of creativity are applied to the following subjects.

1. Cars.
2. Propellants for traveling on land.
3. Propellants for traveling on water.
4. Bicycles.
5. Cooking.
6. Home heating.
7. Faucets.
8. Furniture.
9. Door locks.
10. Sewing machines.
11. Washing machines.
12. Vacuum cleaners.
13. Heat engines.
14. Refrigerators.
15. Radio and TV receivers and transmitters.

II. Propose and create technical solutions for following problems (after finding the solutions, define the complexity level of the problem).

1. Design a device for cleaning the air of tobacco smoke; it may be something that switches on automatically when

the concentration of the smoke reaches a certain level. It should be suitable for offices, trains, and so on.

2. Design a shower actuated automatically when a person walks under the shower head; electrical devices are not allowed because of the danger inherent in a combination of electricity and dampness. The device could be suitable, for example, for public showers at the beach.

3. Propose a design for a hydraulic-pressure amplifier that is fed (for instance) by a water pressure of $P_1 \approx 10$ atm without the use of any other energy source. Such a device might be useful for irrigation.

4. Existing water-pressure regulators include an elastic element whose deformation dictates the control policy. Create a water-pressure regulator that is free of the influence of nonstable elastic elements (its properties change with time).

5. Create an inflatable umbrella. It should have a minimum of rigid mechanical parts, and should not interfere with its carrier's entrance into a car, a bus, or a shop, or with passersby.

6. Existing DC electric motors are provided with a commutator or collector that has sliding contacts. This results in losses in torque, and small motors, in particular, suffer from this problem. Find a solution that avoids the torque losses due to mechanical friction of the contacts.

7. Propose a device able to move on walls or, generally speaking, on vertical surfaces.

8. Figure 4E-1 shows a common assembly. The body 1 has to be fixed on the support 2 while rollers 3 must separate the two bodies. The problem is that this procedure has to be accomplished by one worker, who has difficulty in controlling both the body and the rollers. Propose a solution that simplifies this activity.

9. Roller bearings consist of two rings and a special body that separates the rolling elements. Although the separator is not loaded, it causes friction. Propose, for example, a roller bearing that does not have these inherent losses.

10. Night driving involves the driver having to switch the bright lights of the car on and off. It causes inconvenience to all drivers on the road. Propose a way to decrease, or completely avoid, this inconvenience.

11. You have a hungry baby who does not care about your problems and cries at night, disturbing your rest. Find a harmless solution that will, at leaast partially, allow you to get a night's rest.

12. Propose a "telltale" device to be installed in a car that will show a traffic inspector that the speed limit has been exceeded. The design of the device must take into account that honesty is not an inherent property of human beings.

13. Design a device that will inform the handlers of a parcel or package that it is suffering undesirable shocks or positioning during its transportation (for instance, a parcel labeled "fragile").

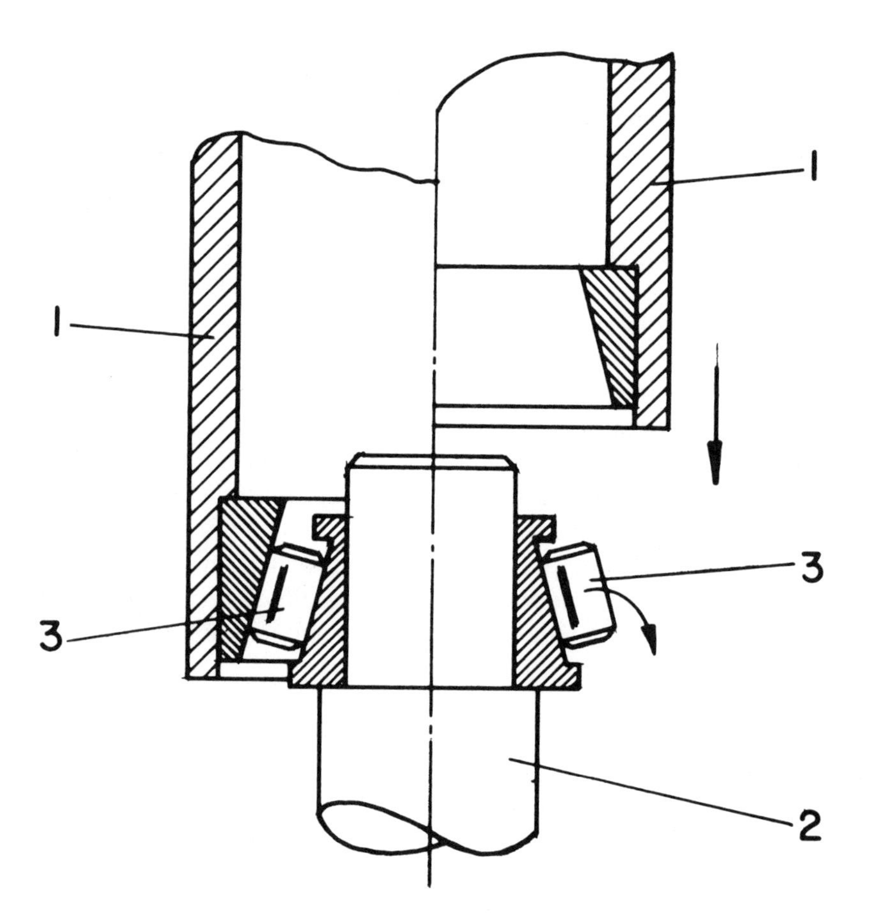

Figure 4E-1 Exercise II-8.

14. Design an automatic bathroom lock. When one enters the bathroom, the door is automatically locked and can be opened only from inside. After one leaves the bathroom, the door is not locked.

15. Propose a way of utilizing the wave energy of an open body of water to charge accumulators for illumination of a buoy or for its radio transmitter.

16. Propose a solution to the problem of measuring small speed oscillations in a rapidly rotating shaft.

17. When approaching the resonance speed a rapidly rotating shaft begins to vibrate. Try to find a solution that provides "smooth" working conditions regardless of the rotating speed.

18. In many places the shore is soiled by crude oil spills. Can you propose a device or method for cleaning the sand?

19. Create an accurate automatic mixer for two or more gas components.

20. Working the soil on slopes of hills for agricultural purposes is difficult, especially since the usual type of tractor can overturn on such a slope. What is your proposal?

III. What are the inverse solutions to the following system?

1. The peristaltic pump shown in Figure 4E-2 consists of a housing 1, a rotor 2 provided with two rollers 3, and an elastic pipe 4. When the rotor 2 rotates, the rollers 3 are pressed on the pipe 4 and push out the portion of the liquid enclosed between points A and B (it is possible to use more than two rollers). In this manner a continuous flow of the liquid through the pipe is produced.

2. The magnetohydrodynamic device shown in Figure 4E-3 consists of a trough containing seawater 1 connected to a pair of electrodes 2. A magnetic field is produced by means of magnets 3. The interaction of two fields causes the flow of water.

IV. Propose technical solutions by combining the following elements (Principle III).

1. Pneumatics with hydraulics.
2. Cam mechanism with gearing.
3. Cam mechanism with a linkage.

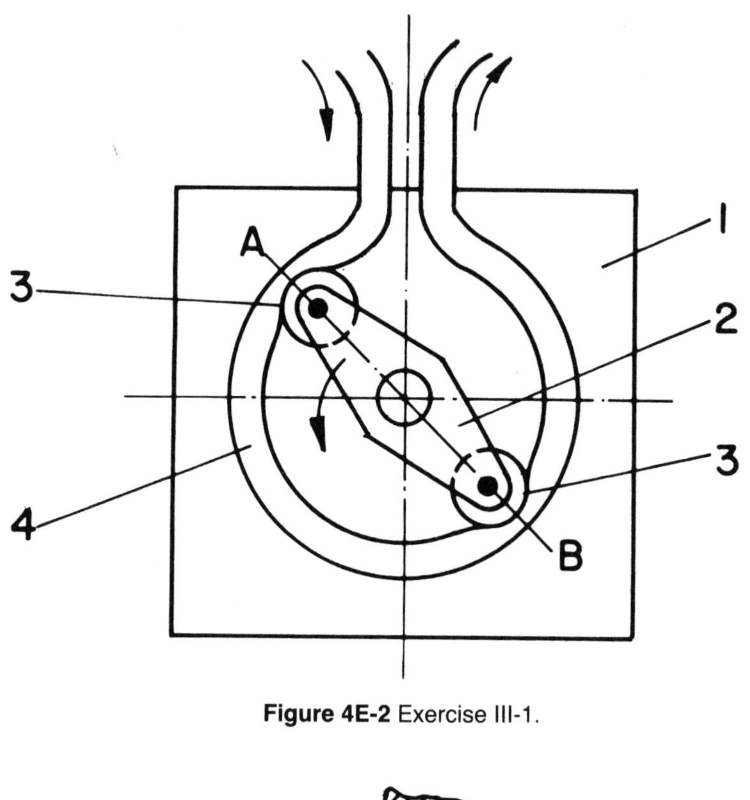

Figure 4E-2 Exercise III-1.

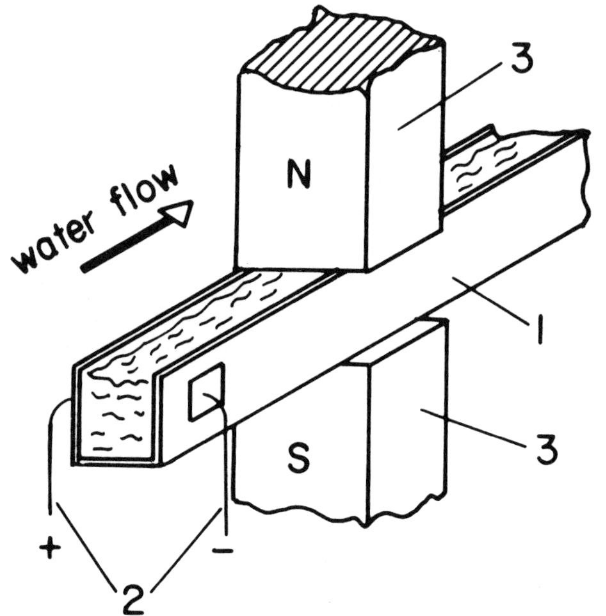

Figure 4E-3 Exercise III-2.

PART II

INTRODUCTION

The engineering design process must begin with a clear definition of the action the object to be designed has to perform. An excellent description of this aim is given in Edward V. Krick's book *An Introduction to Engineering and Engineering Design:* "Each problem can be formulated as a change of the state *A* into a state *B*." Each problem takes the form of a "black box" in which state *A* exists at the input and state *B* at the output. A number of examples illustrating this statement are given in Figure II-1. The "what" problem is solved by defining the nature of the states *A* and *B*, and the "how" problem is embodied in the black box.

For example, the 300- to 400-year-old desire to transmute fuel into mechanical torque culminated in the creation of the engine, with the type of engine depending on the kind of fuel and the thermodynamic cycle. Similarly, the path from flour, water, and eggs to bread might be described as an ancient manual process or as a modern mass-production technique (the product of the latter perhaps being less tasty). The path itself depends on which "how" is embodied in the black box.

Thus, in brief, after *A* and *B* have been defined, the designer has to "unravel" the contents of the black box. We consider the black box to contain, in general, three stages of design:

1. The design of the "processing" layout.
2. The design of the "kinematic" layout.
3. The design of the structure.

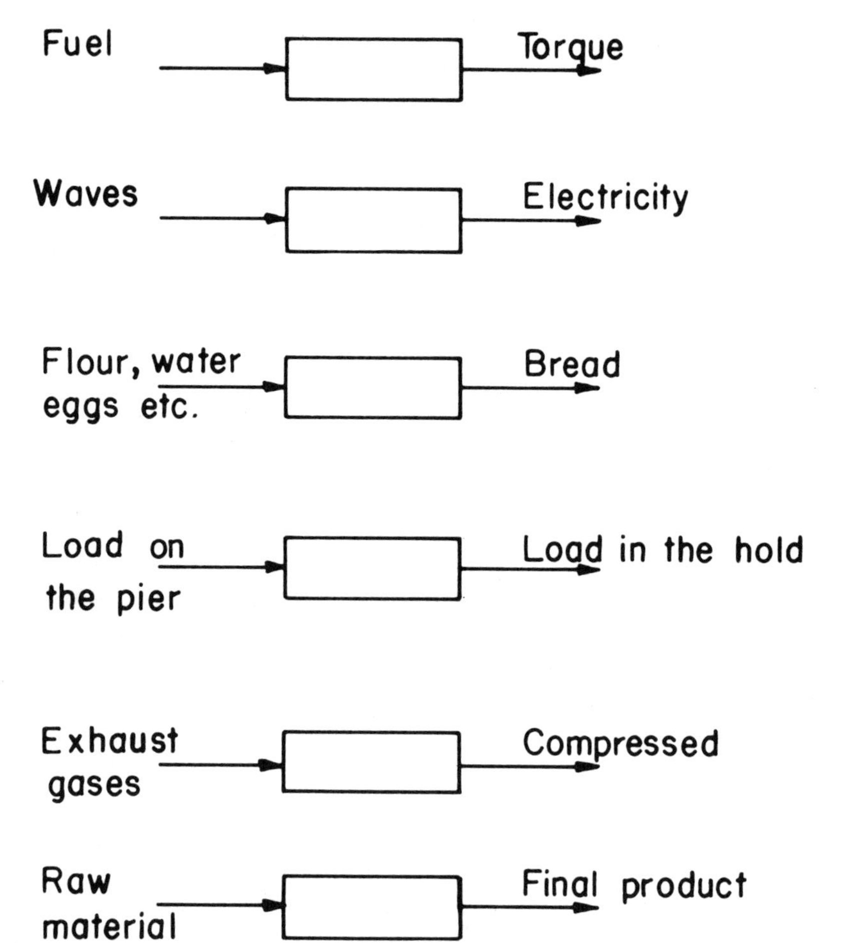

Figure II-1 Each design problem can be formulated as a change of the state *A* into a state *B*.

Chapter 

DESIGN OF THE PROCESSING LAYOUT

During this stage the designer has to put on paper in a schematic way the sequence of processes the black box must carry out, the means or tools involved in these processes, and the main movements that the tools produce. At this stage kinematic (i.e., the transmission of movement) considerations are not invoked. Similarly, the designer does not consider how to obtain the sequence and type of displacements of tools and other elements in the proposed processing layout. The following examples will serve to clarify the foregoing statement.

EXAMPLE 1—A CRANE

Stage A: The load is on the pier.

Stage B: The load is in the hold of a ship.

The processing layout is shown in Figure 5-1. The tool in this case is a hook (for a fork-lift truck, the tool would be a fork). The arrows show the chosen trajectory of the load and the idle stroke when the hook returns to the pier to pick up the next load.

Later we investigate how such a motion can be obtained, but at this stage all we need to know are the dimensions or ranges of the hook displacement in all directions, the allowed and desired

lifting and horizontal speeds of the traveling load, and the parameters of the load to be moved by means of the crane.

EXAMPLE 2—AN INTERNAL-COMBUSTION ENGINE

State A: A certain fuel.

State B: A particular torque and speed on a rotating shaft.

From Figure 5-2 we can see that the fuel is placed in a special tank 1 from where it is pumped by means of a fuel pump 2 into a mixing system in which the combustion mixture is prepared. A carburetor serves as the mixing system and consists of a fuel-splashing device and an air channel. The air is cleaned by passing it through a filter. To create the air flow, a piston moves down inside a cylinder. As soon as the combustion mixture has been

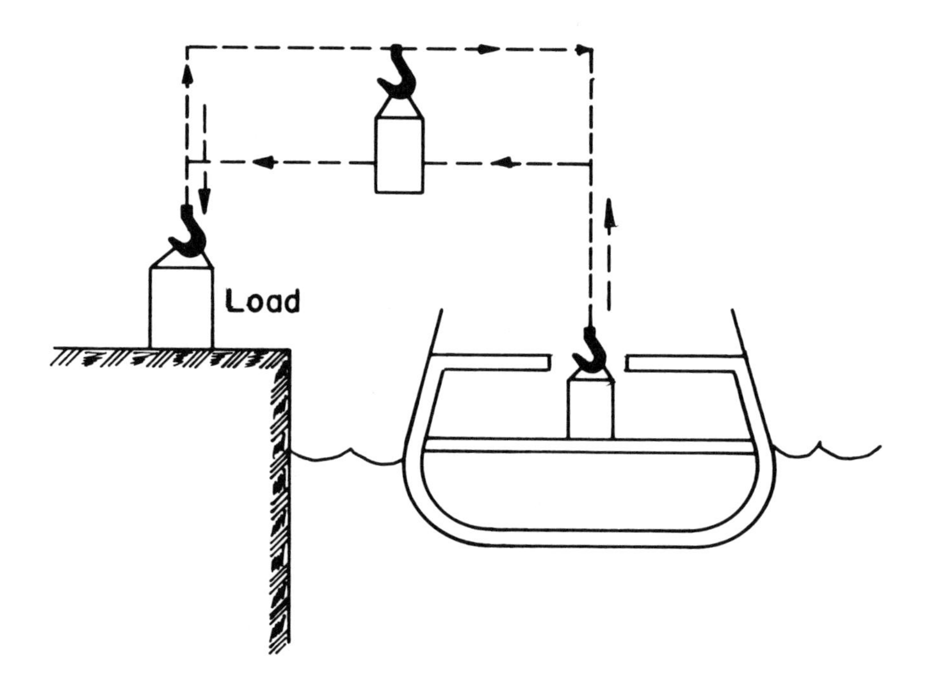

Figure 5-1 Processing layout for a case in which: state *A*—the load is on the pier; state *B*—the load is in the hold of a ship

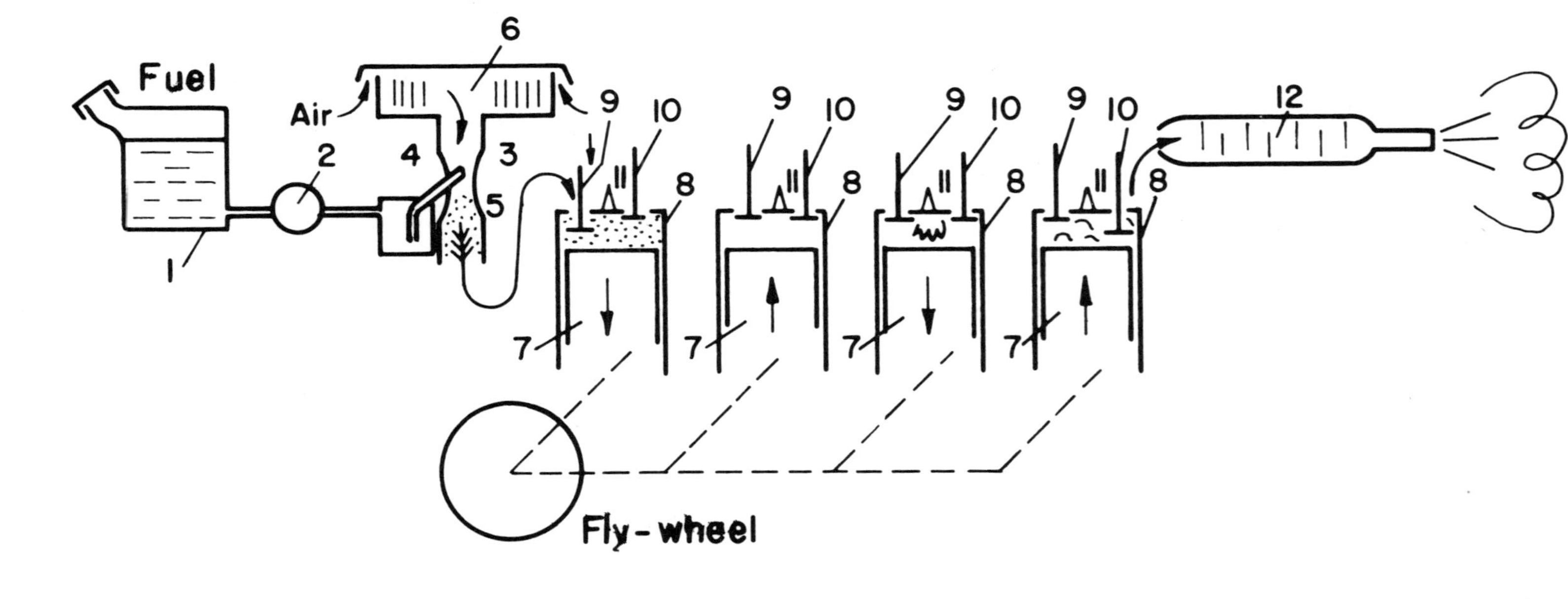

Figure 5-2 Processing layout for a case in which: state *A*—a certain fuel is supplied; stage *B*—a particular speed and torque are obtained on a rotating shaft.

prepared, the inlet valve is opened and the mixture is sucked into the combustion chamber. The space created between the cylinder walls and the top of the piston, which will serve as the place where the mixture is burned, we call the combustion chamber.

In accordance with the desired thermodynamic cycle, the next stroke of the piston must compress the mixture in the combustion chamber. Thus the piston moves upward and the valves are closed. When the pressure in the chamber reaches about 10 atm, the critical point of combustion has been reached, and an electric spark from the plug ignites the mixture. The working stroke thus begins, and it pushes the piston downward. A fourth stroke is used to empty the combustion chamber. During this stroke the piston moves upward and pushes the exhaust gases via the open outlet valve into the noise extinguisher and from there out to pollute the air. To guarantee continuous movement of the piston during the four strokes, the use of a flywheel has been proposed. This device is somehow driven (at this stage we do not think about how) by a mechanism converting the reciprocating displacement of the piston into the rotation of the flywheel.

EXAMPLE 3—THE WEAVING PROCESS

Stage A: Yarn.

Stage B: A fabric.

This process is almost as old as the history of humankind. The layout shown in Figure 5-3 has not changed over the centuries—from Penelope to modern times. The warp (the lengthwise threads) is divided into two (not necessarily equal) sets of threads, *A* and *B*, in such a way as to generate a clearance gap. A special body called a shuttle, which is provided with yarn from a coil, is moved through the gaps from *A* to *B* to *A*, and so on, thus generating the weft. A comblike reed pushes the weft forward, thus thickening the fabric. Thereafter sets *A* and *B* change places, and the process is repeated. The fabric formed is wound onto a coil.

Let us consider ways of moving the shuttle. It can be moved manually, as was done by Penelope, or mechanically, as shown in

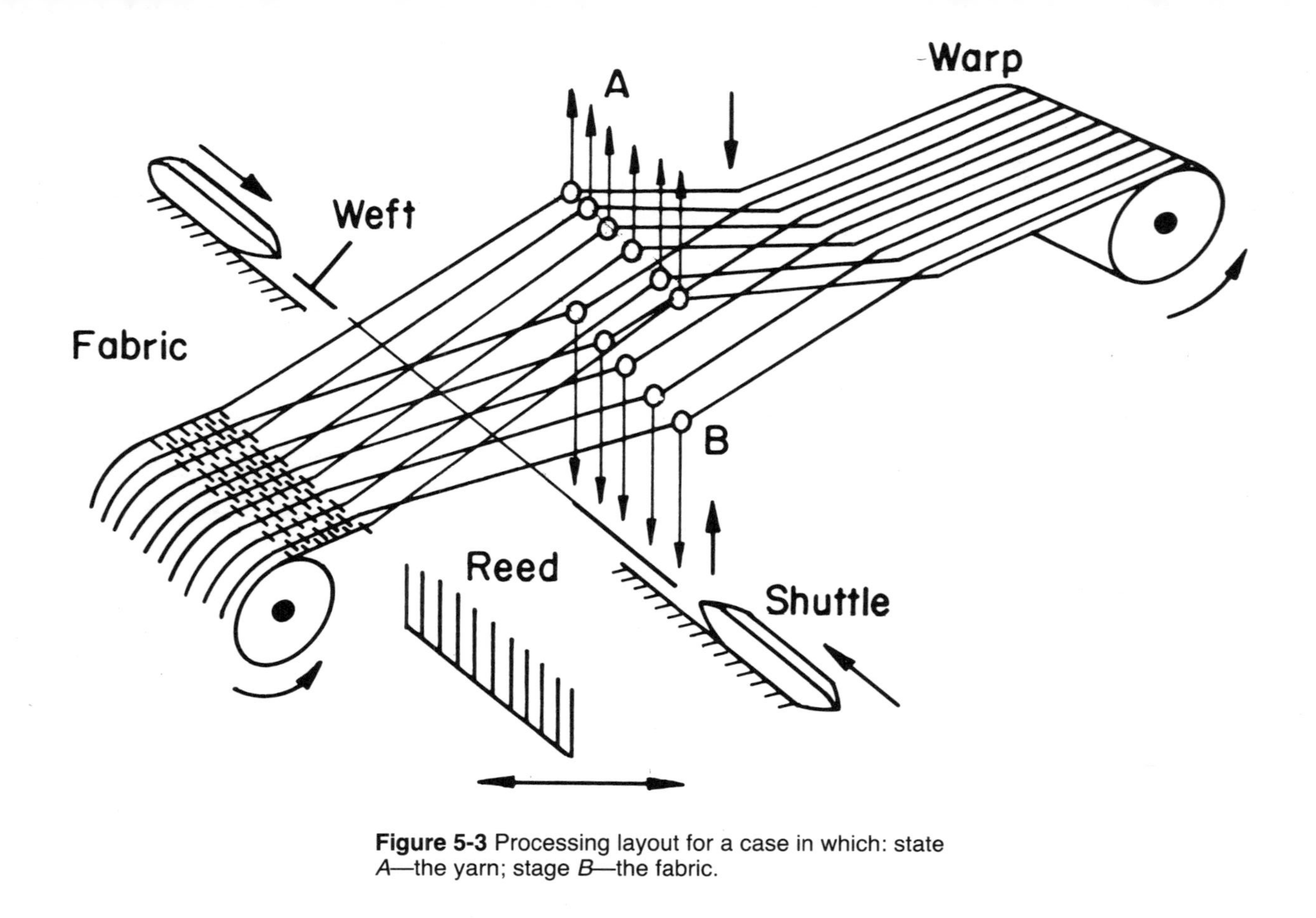

Figure 5-3 Processing layout for a case in which: state *A*—the yarn; stage *B*—the fabric.

Figure 5-4. In case I, the most common case, a mechanical "arm" hurls the shuttle through the gap between the two sets of threads constituting the warp (sometimes over a distance of 7–10 meters). In Case II a "pneumatic gun" shoots the shuttle, and in case III an "electromagnetic gun" serves the same purpose. A different mechanical solution is given in case IV—two permanently rotating rollers "clasp" the shuttle, and generate the initial speed required to throw it.

An excellent example of Principle IV is shown in Figure 5-4V. Instead of a "pneumatic gun" to shoot the shuttle, a "gun" to shoot the yarn was created, and looms that do not use shuttles were thus built. The advantage of this solution lies in the considerable reduction of the mass with which the mechanism must deal.

EXAMPLE 4—THE WIRE-DRAWING PROCESS

This example is illustrated in Figure 5-5. The wire is pulled, by means of rollers, through a series of drawplates, and in this way its diameter is reduced. Of course, as the rotation speed of the roller increases, the diameter of the wire is decreased. The essential difference between this and the three previous examples is that wiredrawing is a continuous nonperiodic operation whereas the other processes are periodic.

It is important to note that the same "processing" concept may be realized in a number of different ways. For example, Figure 5-6 illustrates a metal-cutting process carried out in five different ways, depending on the tools employed and their movements.

A continuous nonperiodic process is preferable to a periodic operation, since there is no waste of time in a continuous process; each working moment is exploited. During the wire-drawing process, for example, wire is produced continuously, whereas during weaving fabric is produced step by step. Here each operation is followed by the next, and only at the end of a certain period can another piece of fabric be considered ready. The periodicity of the weaving process is the result of the actual sequence of operations: the weft has a limited length, the sets of yarn in the warp have to move to two defined, final positions, and the reed can act

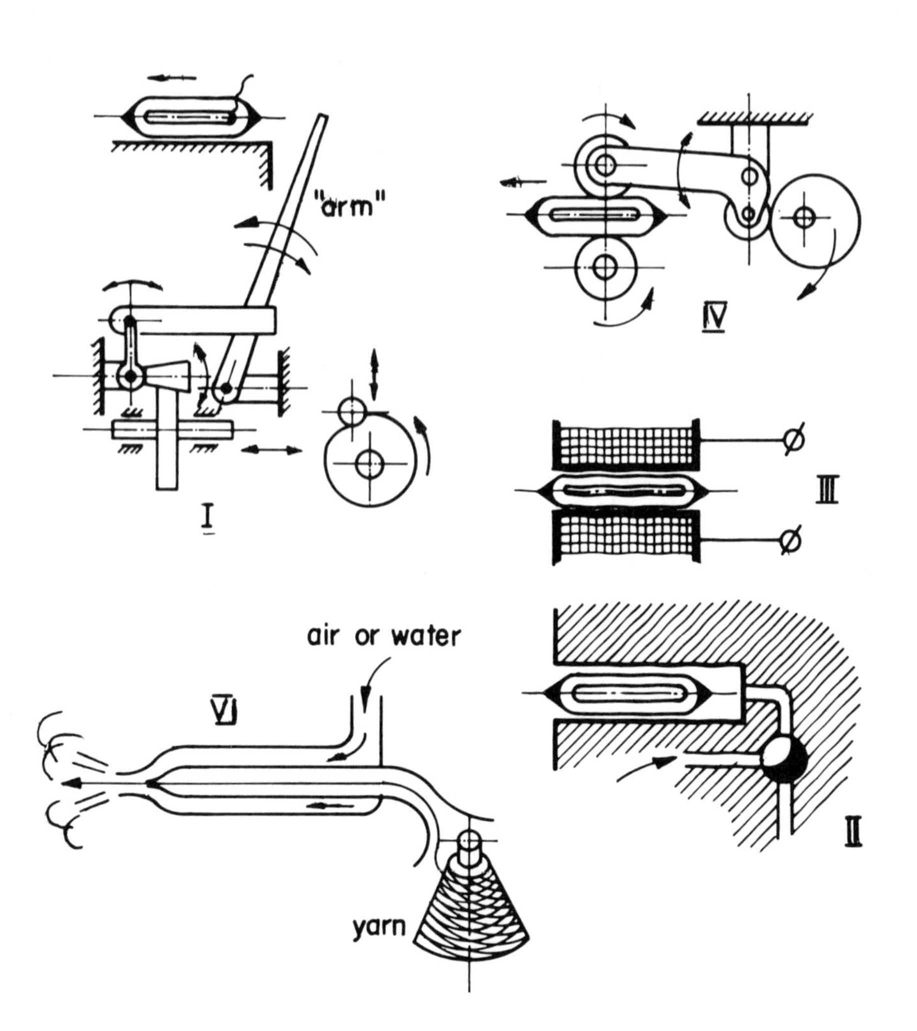

Figure 5-4 Different ways of moving the shuttle of a loom:
I—Conventional mechanical drive.
II—Pneumatic "gun" to shoot the shuttle.
III—Electromagnetic "gun" to shoot the shuttle.
IV—Frictional drive of the shuttle.
V—Ejection of the weft without any shuttle.

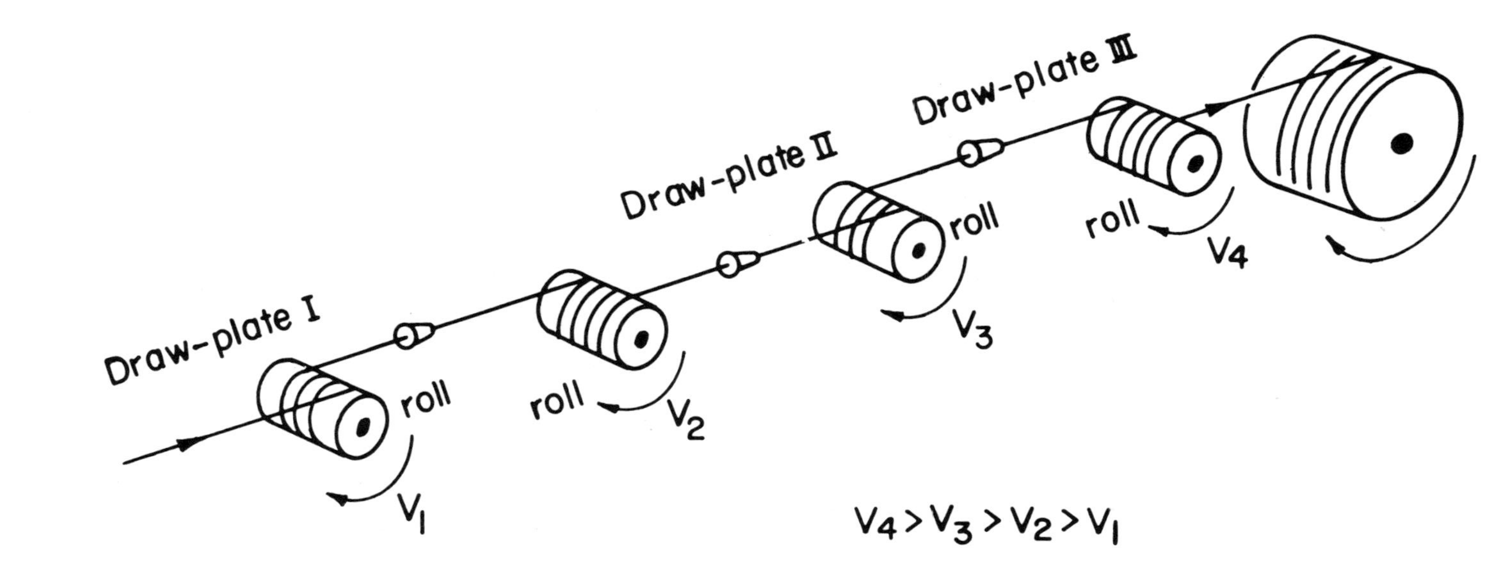

Figure 5-5 Processing layout of a wire-drawing machine. This is an example of a continuous process whereas the other processes presented are periodic.

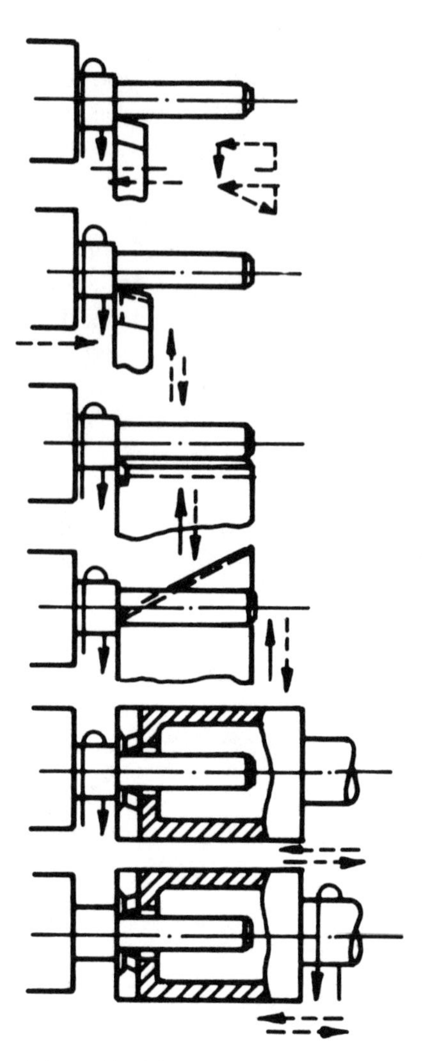

Figure 5-6 The same "processing" concept (in this case metal cutting) may be realized in a number of different ways, as shown (from top to bottom):

(1) The part rotates, and the cutter executes a two-dimensional movement.

(2) The part rotates and moves in an axial direction and the cutter moves only transversally.

(3) The part rotates, and a wide cutter moves only transversally.

(4) The part rotates, and a specially shaped cutter moves only transversally.

(5) The part rotates, and a tool-head moves axially.

(6) The part is immovable and a tool head carries out the required movements.

only after the waft has been woven. The weaving process can be transformed into a nonperiodic operation only if these limitations are overcome. An idea for such a process is shown in Figure 5-7. In this design the warp forms a cylinder, as does the finished fabric. A cross section of the warp yarns is shown in the figure, from which it can be seen that the two sets of threads form two overlapping circles. The crossing points *A* and *B* rotate continuously at equal speeds, thus forming gaps in which the continuously moving shuttles are placed, leaving behind them two wefts. In this type of loom, all the operations proceed simultaneously. These looms are used for the production of rough-weave fabrics, such as burlap bags. The limitations on their wider application lie in their kinematics, for example, the difficulty of moving the "surrounded-by-yarn" shuttles.

Return for a moment to Example 2, the internal-combustion engine, and remember that it too has a nonperiodic successor—the

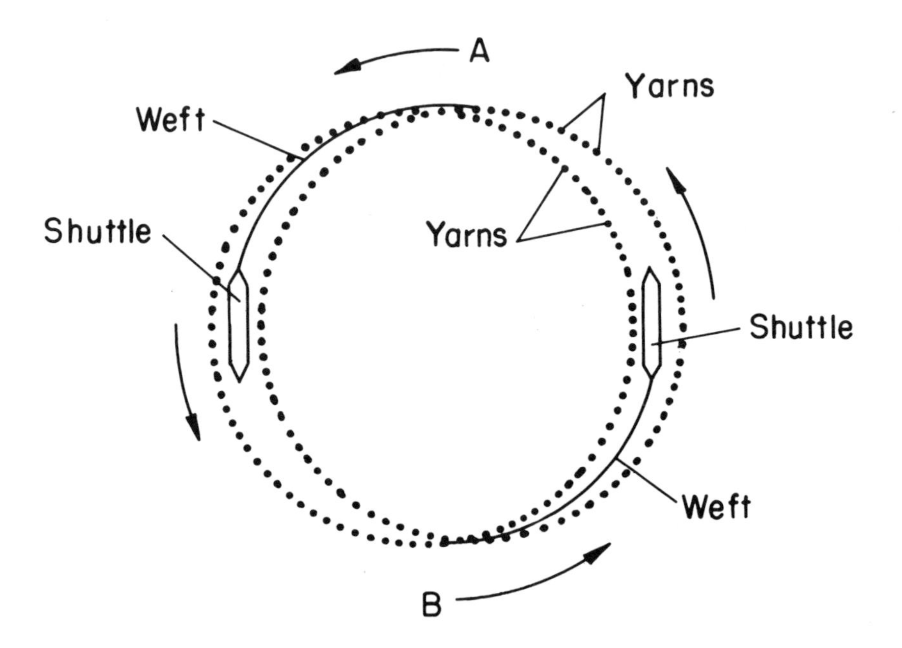

Figure 5-7 An idea for a continuous weaving process. The shuttles move around pulling the weft and the yarns "open" the way for the shuttles.

gas turbine. Thus we see that the chief characteristic of a periodic producing process is a particular sequence of actions for each tool or element. To clarify this sequence, and to make it easy to estimate the time required for each operation, we must design a sequence "program," which we call a "cyclogram."

There are two ways to express a cyclogram on paper: (1) in a cycle form; and (2) in a linear form. The cycle form is convenient because it represents the process and its component parts in direct relation to a rotating body, which we call the "main shaft." In general, most machines do actually contain a main shaft, and the time it takes to complete one or more revolutions defines the period of the process. Figure 5-8 presents a cyclogram of an internal-combustion engine (Otto cycle) according to the layout in Figure 5-2. In this case the main shaft is the crankshaft of the engine, and one period requires two revolutions of the shaft. The rest is clear from the Figure. The angles α_1, α_2, α_3, α_4, and α_5 define,

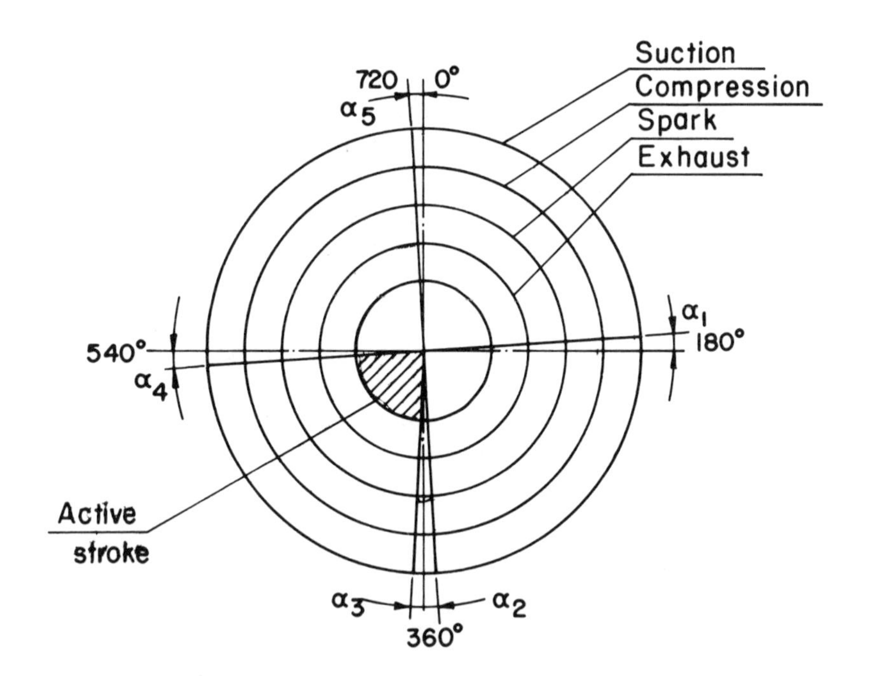

Figure 5-8 Cyclogram of an internal-combustion engine (Otto cycle).

with a high degree of accuracy, the opening and closing moments of the valves and the time at which the mixture burns. The angles of rotation of the main shaft can easily be translated into time units by knowing the value of the rotating speed (and assuming it to be uniform). Two inherent disadvantages of this cyclic cyclogram are:

1. Although the relative values of the angles are represented in a most obvious manner, the absolute lengths of the arcs that represent these angles depend on the radius of the corresponding circle. This fact sometimes has a psychological influence on the "reader" of the cyclogram.
2. The course of the motion or action of each tool or element cannot be fully described; only the times at the beginning and end of the action can be pinpointed. In the case of the internal-combustion engine, for example, the displacements of the valves as functions of time cannot be seen.

These disadvantages can be avoided by the use of a linear cyclogram. In Figure 5-9 a cyclogram of the same engine is presented, but this time in linear form. Note the graphical interpretation of the valve motion. Here it is possible to show the maximum displacement of any valve by using the vertical coordinate and a defined scale. (If required, the forms of the acceleration and deceleration can also be shown.) Obviously the spark action does not require the use of this coordinate.

The vertical coordinate can also be related to other physical values, such as pressure, temperature, and voltage, if the action being considered requires it, and the form of the changes in these physical values can appear on the linear cyclogram. These cyclograms show:

1. The sequence of action.
2. The relationships between the time intervals required for each action.
3. The time reserves that may help to decrease that period.
4. The "shapes" of the movements or other changes in various parameters as functions of time.
5. The initial information needed for the design of the kinematic layout.

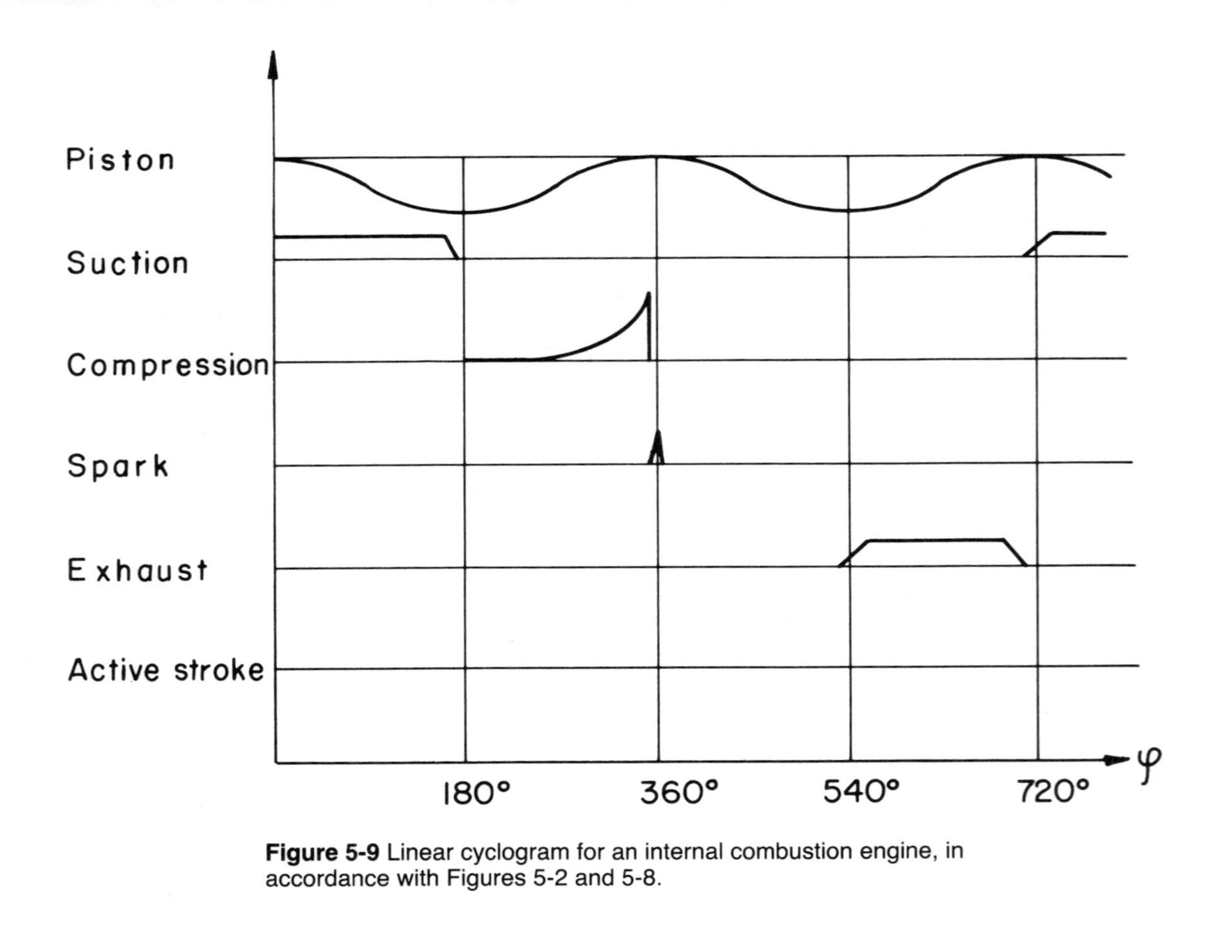

Figure 5-9 Linear cyclogram for an internal combustion engine, in accordance with Figures 5-2 and 5-8.

At this stage the conceptual design once again comes into contact with engineering calculations. The designer has to estimate the operation times of each of the elements regardless of the nature of the process under consideration. Both the designer's knowledge of mathematics, physics, mechanics, hydraulics, and so on, and the designer's skill determine the quality of the design. This is the point at which the limitations described when design principles were discussed will become evident.

Let us now try to design a processing layout—for example, a process for the automatic manufacture of an aluminum antenna, as in Figure 5-10. Since about half a million antennas are manufactured each year, it is natural that the raw material should be in the form of a coil of aluminum wire.

The coil of wire 1 is placed on a freely rotating turret 2 and is unwound by a pulling force, as shown in Figure 5-11 (position *a*). The unwound wire, which is now bent into a spiral form, must be straightened out before anything can be made from it, and for this purpose a classic "straightening" tool is used, as follows: The wire is passed through a rapidly rotating bushing 4 fitted with five pins 3 (as shown in Figure 5-11, position *b*), in which it undergoes plastic deformation. The straightened wire is pulled out of the bushing by two rotating rolls, which create the required pulling force by means of friction (Figure 5-11, position *c*). The next step is the measuring off of the length of wire needed to produce a single antenna, about 1186 ± 5 mm. A high degree of accuracy is not required, and a simple electric contact 6 placed in the path of the running wire (Figure 5-11, position *d*) can be used. This contact has a dual purpose: it stops the pulling and straightening devices and activates the cutting and bending tools. The measured section of the wire is then cut off (Figure 5-11, position *e*) by means of a cutter moving upward. The piece of wire is fixed in a horizontal position by some means (see horizontal view) and the bending operations can begin. Since the antenna is symmetric, the bending operations on each side of it can be carried out simultaneously, thus saving time.

The first operation is the bending of the two ends to form two vertical 40 ± 5-mm sections. This is achieved by moving the tools 8 upward and bending the wire around two fixed pins 9 (Figure 5-11, position *f*).

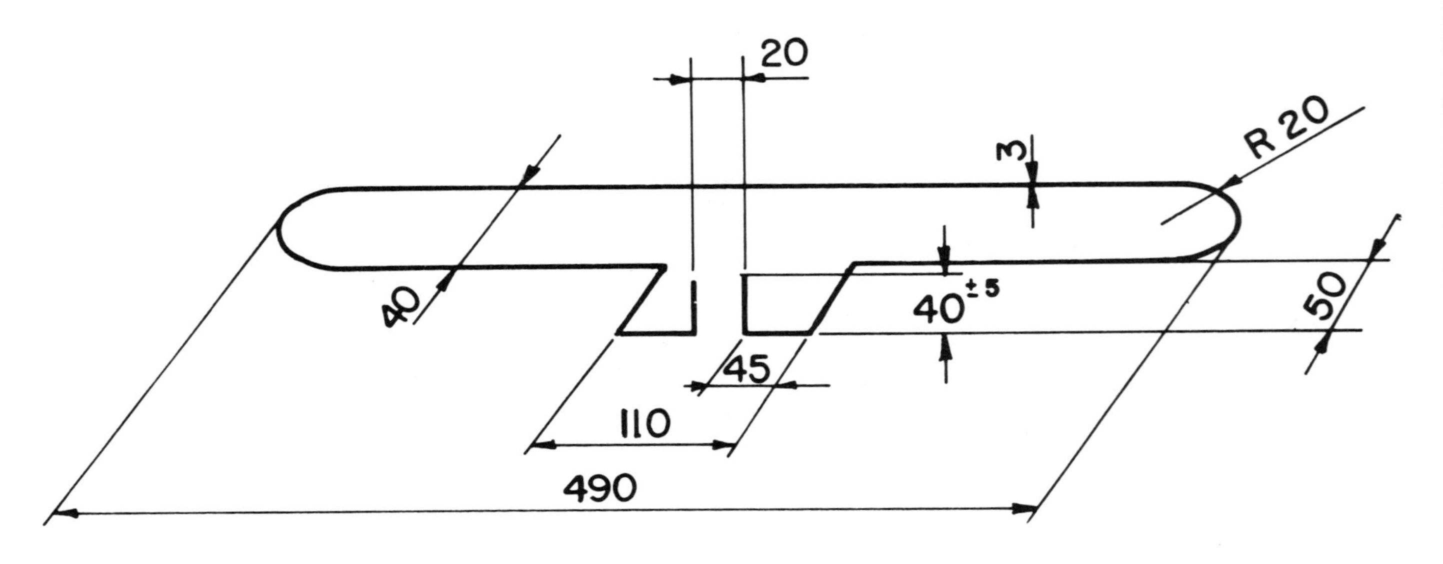

Figure 5-10 An aluminum antenna. The design process of an automatic machine for manufacturing this type of antenna is shown in Figures 5-11 and 5-12.

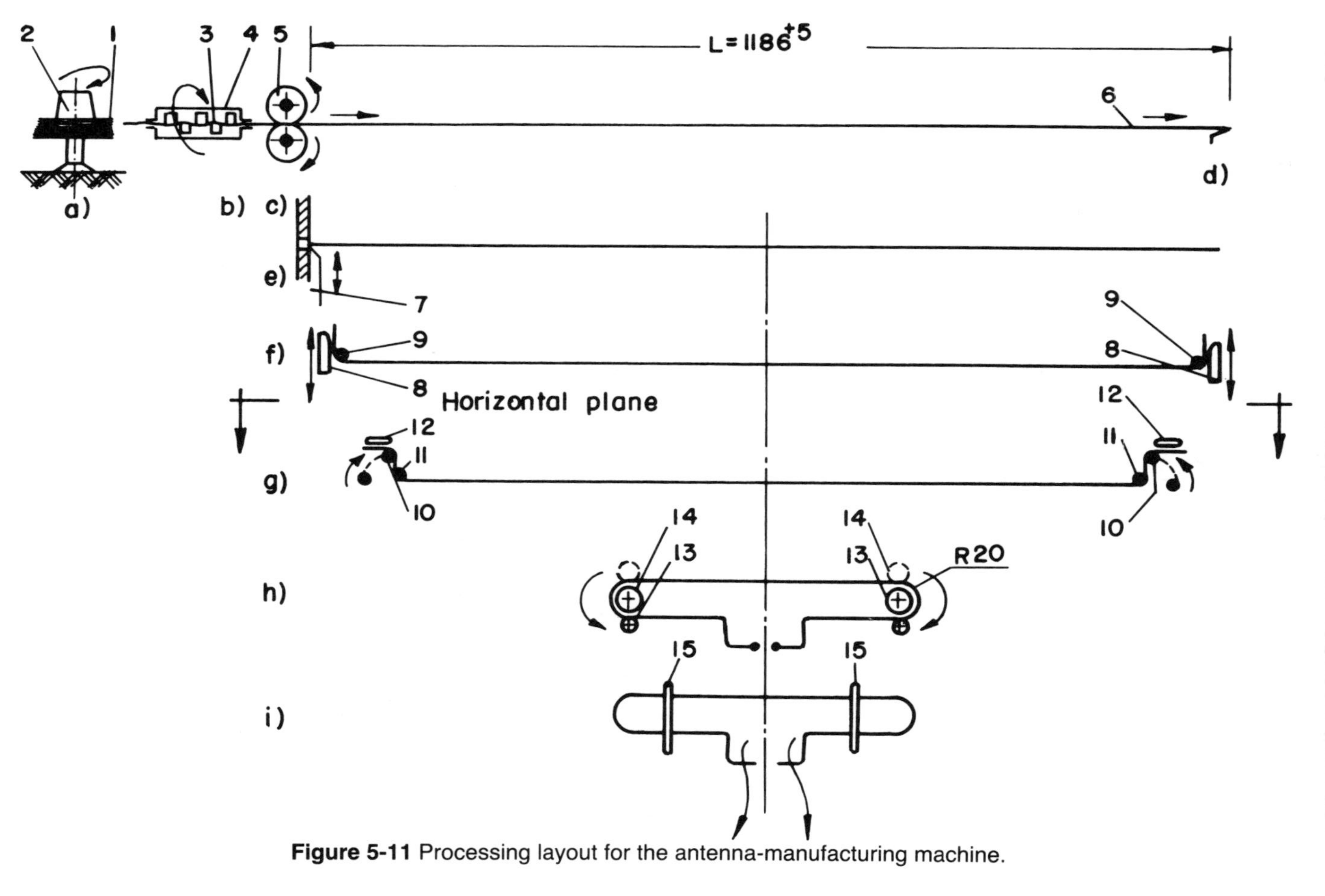

Figure 5-11 Processing layout for the antenna-manufacturing machine.

The subsequent bending operations are carried out in the horizontal plane and thus all the diagrams are drawn in that plane. The bending of the 45-mm and 50-mm sections into a steplike configuration at each end of the wire can be achieved in a single operation, as follows (Figure 5-11, position *g*): On each side a rotating pin 10 bends the wire around a fixed pin 11 to meet a fixed support 12, and in this way the steplike section is completed. The last bending operation, the formation of the curved sections with radius R20, is achieved by bending the wire around two fixed cylinders 14 by means of rollers 13 (Figure 5-11, position *h*).

The final operation (which we call extraction) is the removal of the completed article from the machine. In the layout this step is illustrated in the vertical plane. A lifting body 15 raises the new antenna and it slides into a packing case (Figure 5-11, position *i*).

Figure 5-12 shows a linear cyclogram for this process. Each line corresponds to one of the operations described. In general each action of any tool has three phases: an increase in movement, a uniform phase, and a slowing down. For instance, under the influence of the pulling mechanism (whatever it may be), the motion of the wire first increases (θ_1), then there is an interval of uniform pulling (θ_2), and finally, after the wire has been measured, the mechanism slows down (θ_3). However, sometimes the action of a tool consists of only two phases: increase in movement and slowing down. In our case, for instance, the bending operations do not involve a uniform-motion phase, and a rest period is often needed between the forward and backward strokes.

The process has to be designed in such a way that predetermined time intervals between operations will avoid interference by the different phases with one another.

The dashed lines in Figure 5-12 show a more efficient use of the same process. The period is shortened by eliminating the time intervals wherever consecutive operations, or parts of them, can be carried out simultaneously. For instance, the first bending operation can begin before the cutter has completely returned to its initial position. Similarly, the wire-pulling and -measuring process can begin before the "extraction" tool returns to its initial position. Once again, to know how much time can be saved, we have to move from a conceptual approach to an engineering and computational one. In our case the initial value of the period T equals:

$$T = t_i + t_2 + t_3 + t_4 + t_5 + t_6 + 5\delta \quad 7 \text{ seconds}$$

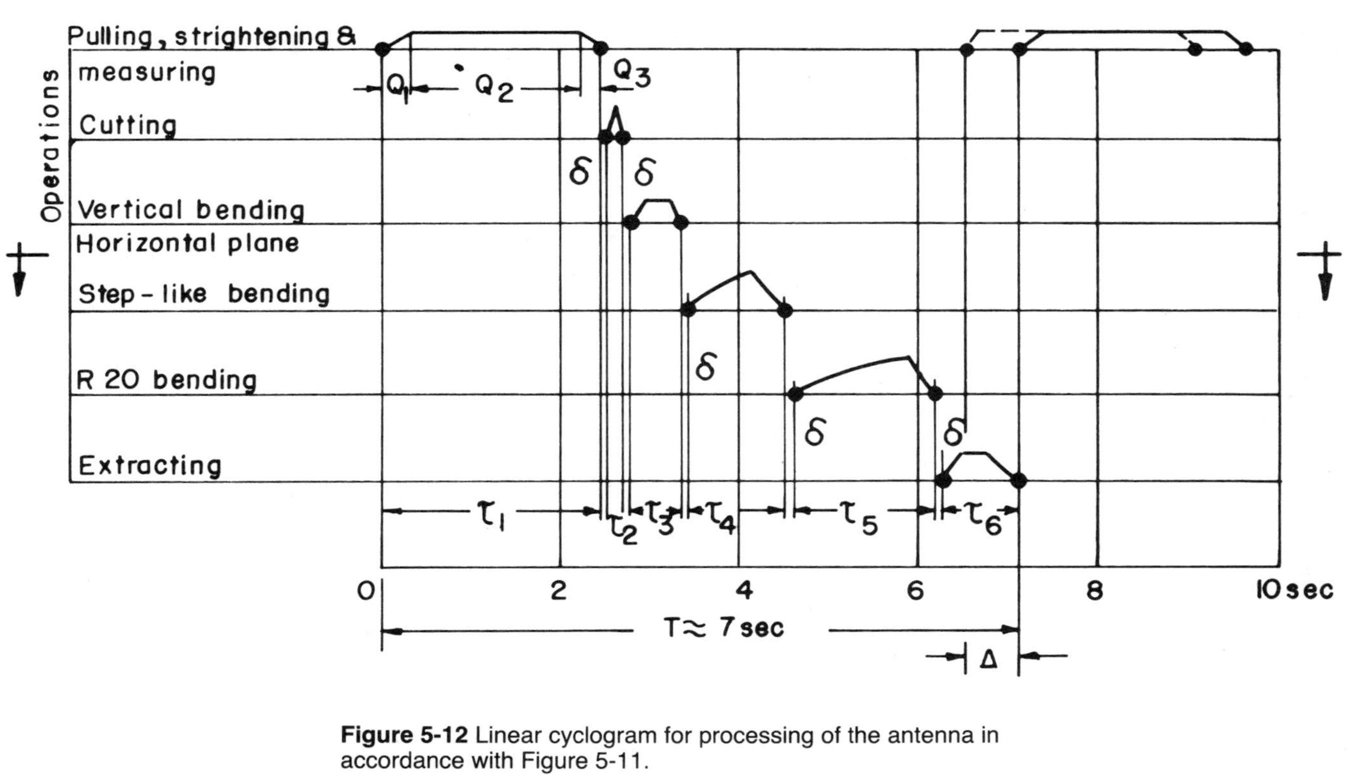

Figure 5-12 Linear cyclogram for processing of the antenna in accordance with Figure 5-11.

The improved value of $T' = T - \Delta = 6.5$ seconds.

While working as a design engineer, this writer was requested to design an automatic, highly productive, machine for manufacturing silver contact needles (Figure 5-13). To understand better the difficulties facing the design team, we have to keep in mind that silver is a very soft material. Thus grinding, for instance, is not a suitable treatment because the silver clings to the grinding stone and blocks its pores. The same problem occurred with the file, the tool used in the manual production of the needle. Consequently our first idea—to borrow from the manual process—failed. It is difficult to imagine how any machine, after the production of each batch of 30 pieces, could clean the file with a metal brush, as did the worker who produced the silver contacts manually. An attempt to treat the contact by cutting—a manufacturing principle borrowed from the metal-turning process—also failed, because of the small dimensions of the part and its inherent softness, which led to bending of the silver wire. We then set about designing a process using the "systematic approach."

- What is the aim?

To produce a part from a particular material in accordance with the given specifications.

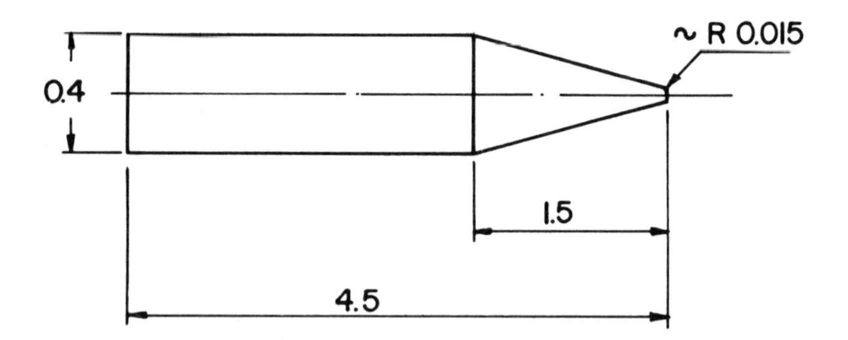

Figure 5-13 Diagram of a silver contact needle. A processing layout and a tool for sharpening the needle are required. (The other operations, such as measuring and cutting, are trivial.)

• What are the obstacles?
The properties of the material and the small dimensions of the part.
• How are the obstacles manifested in the processing?
Grinding and filing are not acceptable because of sticking, and cutting is not possible because of the bending of the wire and the small dimensions of the part.
• What is the way out of the situation?
Perhaps to create a grinding tool with very small pores so that the silver will not clog them.

This last thought led to the creation of the tool shown in Figure 5-14. It consists of two disklike bodies assembled, as is shown in the figure, in such a way that the triangular slot so formed shapes the tip of the needle. To avoid bending the wire, we decided that it should be rotated so as not "to allow time" for curving. A machine rotates the wire at about 600 r/min and pushes it into the slot of the disks, which, in turn, rotate at about 1500 r/min. This method proved very successful. The other operations (feeding and cutting) are trivial.

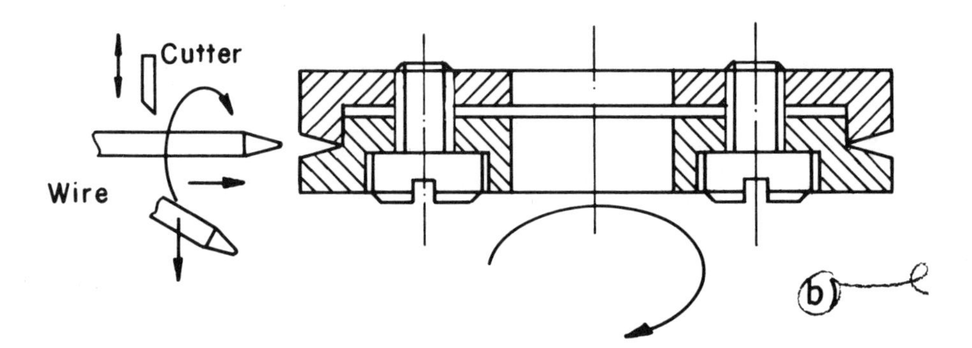

Figure 5-14 The silver needle in Figure 5-13 was sharpened by a rapidly rotating hardened steel tool of the shape shown here.

EXERCISES

What would be your proposed processing layout for each of the following?

1. Automatic or semiautomatic manufacturing of a chain corresponding to Figure 5E-1.
2. Automatic manufacturing of a spiral spring in accordance with
   a. Figure 5E-2a.
   b. Figure 5E-2b.
   c. Figure 5E-2c.
3. Automatic wrapping of bars of soap.
4. Automatic gluing of labels on cans or bottles.
5. Automatic packaging of eggs in boxes.
6. Automatic or semiautomatic conductor winding on a ferrit ring for:
   a. A large number of turns.
   b. A small number of turns.

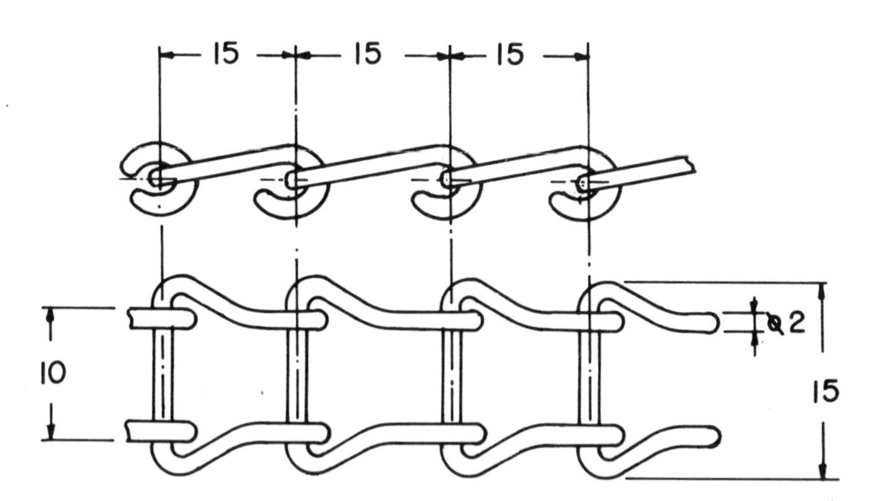

Figure 5E-1 Exercise 1. How would you design a processing layout for automatic or semiautomatic manufacturing of the chain shown?

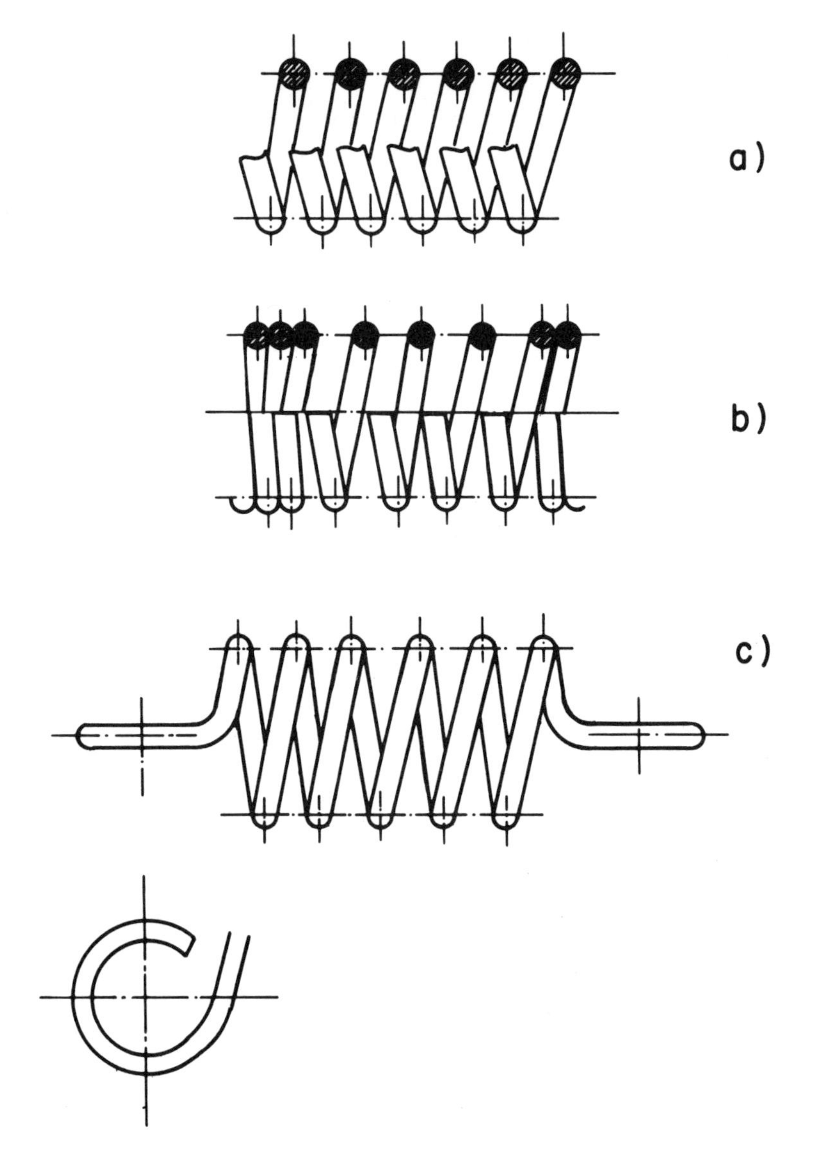

Figure 5E-2 Exercise 2. How would you design a processing layout for automatically manufacturing the spiral springs shown in a), b), and c)?

Chapter 6

KINEMATICS

At this stage of the design, the designer has to find ways to produce the desired movements of the tools, or other working parts, and determining their significance according to the "processing" layout. At the designer's disposal are mechanical, pneumatic, hydraulic, and electrical methods. To make the right choice, one must know the main characteristics of these techniques, and their comparative advantages and disadvantages. A combination of two or more of the four methods may provide a suitable solution, but this must be justified by sound reasoning. Too many communication lines and auxiliary accessories will be required by a "mixed" solution and the expense must be balanced by a definite benefit.

This stage of the kinematic design can be considered to be complete when a layout of the kinematic solution is ready. Of course, a large number of calculations (ratios, and numbers of teeth, belt drive wheels, cylinders, etc.) will be involved in the work, but we do not consider these here. Our discussion is limited to concepts and methods of invoking them.

The layout must include all the necessary elements—wheels, shafts, bearings, supports, cylinders, valves, switches, and so forth. The single aim at this stage is the clarification of the kinematic solution. The designer does not have to bother with scales, locations, or proportions.

THE VARIOUS SYSTEMS

The advantages and disadvantages of the various systems are given in the following.

Mechanical Systems

Advantages

1. The relative obviousness of mechanical layouts (in comparison with an electronic circuit, for instance).
2. The absence of the need for a specific power supply.
3. The possibility of implementing different kinds of movements or different displacement laws.
4. The relative ease of achieving accurate displacement.
6. The rigidity of mechanical links.
7. The high accuracy of ratios in movement transmission.

Disadvantages

1. For spatially extended constructions, the cumbersomeness of the mechanical kinematic solution.
2. Difficulties in creating relatively rapid movements.
3. Difficulties in creating very large forces.
4. The necessity for special protecting devices to avoid breakage of expensive links.

Hydraulic Systems

Advantages

1. The possibility of achieving very large forces.
2. The possibility of carrying out slow and smooth movements.
3. The relative simplicity of carrying out spatial location of moving elements.
4. The possibility of changing velocities of displacements in a smooth manner.
5. The fact that it is not explosive (pressure sharply drops when the liquid leaks).

Disadvantages

1. The difficulties resulting from the use of high pressures.
2. Mechanical supports or complicated control layout required for accurate displacements.
4. Leakage influence on the pressure inside the system.
5. Variation of the viscosity influenced by temperature changes.

Pneumatic Systems

Advantages

1. The relative ease of providing complicated spatial locations of moving elements (e.g., pipes can be bent into any shape).
2. The relative ease of carrying out rapid movements (dependent on the thermodynamics of gases).
3. The relative ease of creating large forces (which are the product of the pressure and the area of the piston or diaphragm).

Disadvantages

1. The difficulties in creating special displacement laws.
2. The need for mechanical supports for obtaining accurate displacement.
3. The dependence of the action on the pressure in the piping.
4. The need for special auxiliary equipment.
5. The need for means to avoid leakage.
6. The danger of explosion.

Electrical Systems

Advantages

1. Spatial locations of working elements easily achieved.
2. High rate of automation easily obtained.

Disadvantages

1. Problems of reliability.
2. The need for relatively well-educated maintenance personnel.

We shall illustrate the design of a kinematic layout using the automatic machine for manufacturing an antenna (Figure 5-10) as our example. As with the processing layout, the first stage of the kinematic layout must provide for the straightening of the wire and its simultaneous feeding into a bending position. These two simultaneous processes must be stopped when the wire has reached the required length (about 1186 mm), and remain motionless during the bending process. For this purpose a mechanical approach is the means of choice, the driving being carried out by an electromotor with a rotation speed of about 1500 r/min.

Figure 6-1 shows this part of the kinematic layout. The motor

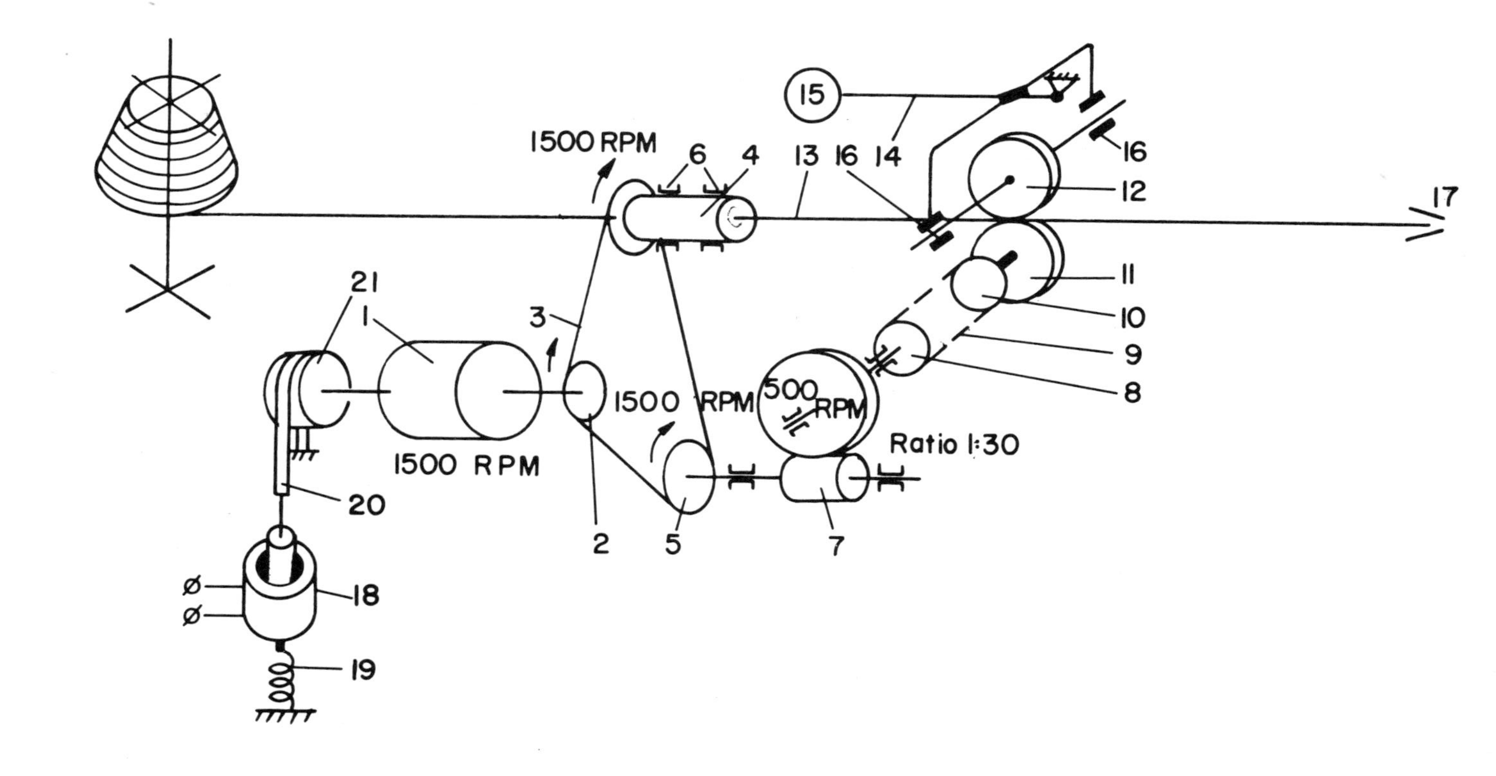

Figure 6-1 Kinematic layout of the antenna-processing machine, showing the mechanical portion.

1, via pulley 2 and trapezoidal belt 3, drives pulleys 4 and 5, pulley 4 being fixed to the straightening device placed on bearings 6. The ratio between the motor and this device is about 1:1. The pulley 5 drives the worm speed reducer 7, which has a chain wheel 8 on its outlet shaft. The chain 9 connects wheel 8 to the other chain wheel 10. The rotation speed of this wheel is about 30 r/min as a result of the action of reducer 7. The wheel 10 drives the lower feeding roller 11. Friction forces between roller 11, the upper roller 12, and the wire 13 have to be sufficiently large to pull the wire. The value of these forces is controlled by the lever 14 and the load 15. The weight of this load is increased by the lever 14, which, via the bearings 16, presses the shaft of the upper roller 12 onto the wire. When the wire 13 reaches the contact 17, it actuates the electrical layout, which disconnects the motor 1 and the coil 18 of the electromagnetic brake from the electric supply, thus stopping the feeding of the wire. The brake is closed by the spring 19, which pulls the belt 20 around the drum 21. The particulars of the electrical layout are given later.

Knowing the rotation speed of the motor and the desired velocities of the rotation of the straightening device and the pulling rollers (to provide the required feeding speed), we can define the necessary ratio values; that is, the diameters of the belt pulleys, the number of teeth, and so forth.

When the feeding of the wire has been accomplished, the required length of wire is available, and this must be cut so that the bending processes can begin. Because of the considerable distances between the bending positions, we decided to use a pneumatic solution for this purpose (Figure 6-2).

A sequence of pneumocylinders acting in concert with the processing layout is provided by a rotating pneumatic valve 22 driven by a motor 23 and a speed reducer consisting of a worm 24 and a wheel 25. The first action to be carried out is the cutting of the wire. For this purpose a piston 26 placed in a cylinder 21 (position *A*) is actuated by air pressure for the working stroke and by a spring 28 for the idle stroke. The piston rod is connected by a link 29 to a lever 30, onto which a cutter 31 is fastened.

Two identical cylinders (position *B*) are used to provide the first bending operation. A punch 32 is fastened to the end of the piston rod, and the wire is bent around a pin 33.

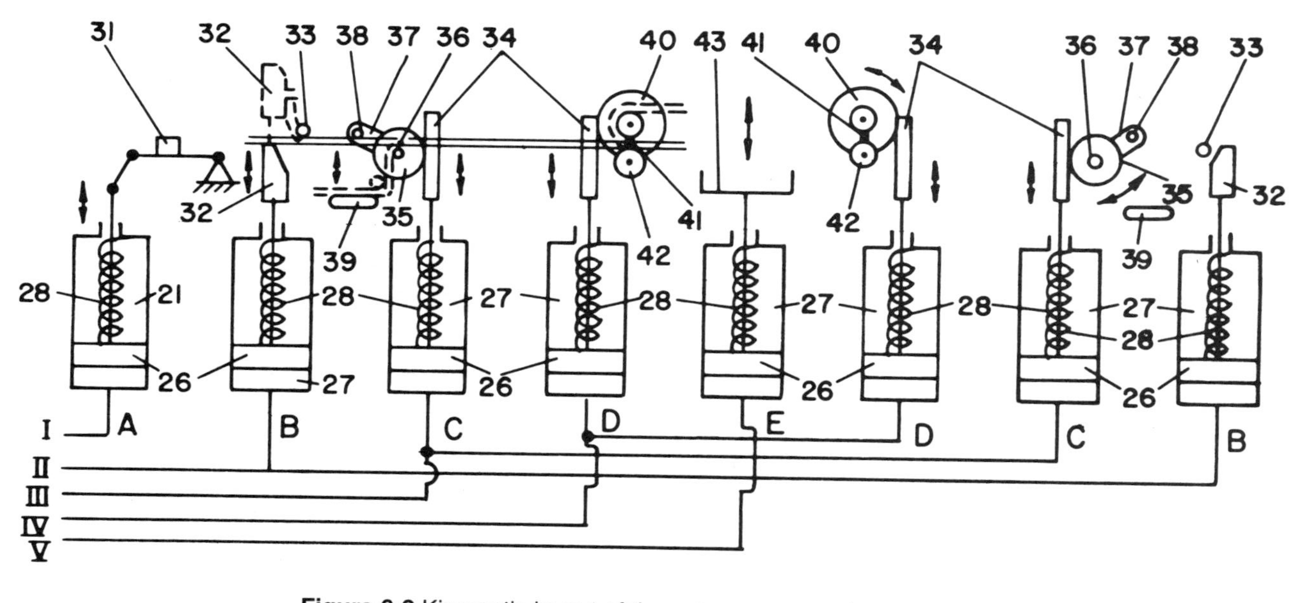

Figure 6-2 Kinematic layout of the antenna-processing machine, showing the pneumatically driven elements.

Another two identical cylinders (position C) perform the second cranklike bending process. A toothed rack 34 constitutes the piston rod, which is geared to a toothed wheel 35 in such a way that the reciprocative motion of the piston 26 is transformed into rotation of a shaft 36. The bending tool 37 is fastened to the end of the shaft and is supplied by two pins 38 and 39. The bending is performed as a result of the rotation of this tool through 90 degrees.

The final bending operation—the formation of a curved section with a radius of 20 mm—is carried out by a third pair of cylinders (position D). Here the piston rod 34 also consists of a toothed rack that transforms the movement of the pistons into the rotation of the wheels 40. The wheels drive a shaft, to the ends of which are fastened another bending tool 41 provided with a roller 42.

The final cylinder (postion E) is used to take the finished product out of the machine. For this purpose a fork 43 is fastened to the end of the piston rod. During the working stroke, the fork raises the antenna and it slides out into a special box, or some other suitable container.

As shown in Figure 6-3, each cylinder is connected by a corresponding pipe I, II, III, IV, or V, to the rotating valve 22 (Figure 6-3). This valve consists of a housing 44 provided with five slots 45, each connected to a corresponding pipe, and a rotating plate

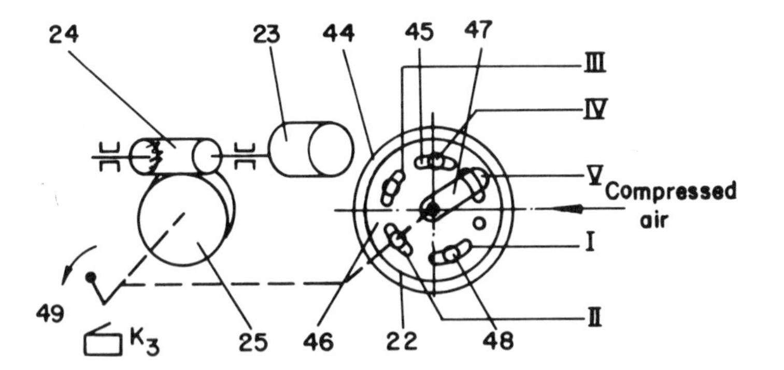

Figure 6-3 Kinematic layout of the pneumatic controller (valve) drive that provides the normal sequence of action of the cylinders (see Figure 6-2).

46 provided with one long slot 47 and five openings 48. The plate 46 is driven by the motor 23. During the rotation the plate 46 connects, by means of the slot 47, the compressed-air source with the corresponding cylinder groups. At that time the other cylinder groups are open, via the openings 48, to the atmosphere, thus allowing the springs 28 to carry out the idle strokes. It is clear that the plate 46 and the housing 44 are well sealed. During the feeding process, the drive of the valve 22 is stopped in such a position that the slot 47 is located in an intermediate sector (i.e., between slots 45) and the cylinders are connected to the atmosphere via openings 48.

The motors are controlled by the electric circuit shown in Figure 6-4. The motors 1 and 23 are three-phase asynchronous machines controlled by contactors K_1 and K_2 respectively. (The contacts of these contactors are denoted by the same letters.)

In the very beginning, at the beginning of the wire-feeding process, the normally closed contacts K_1 in the circuit of motor 1 facilitate the rotation of the motor and actuate the brake's coil 18. This action frees the brake (the electromagnet works against

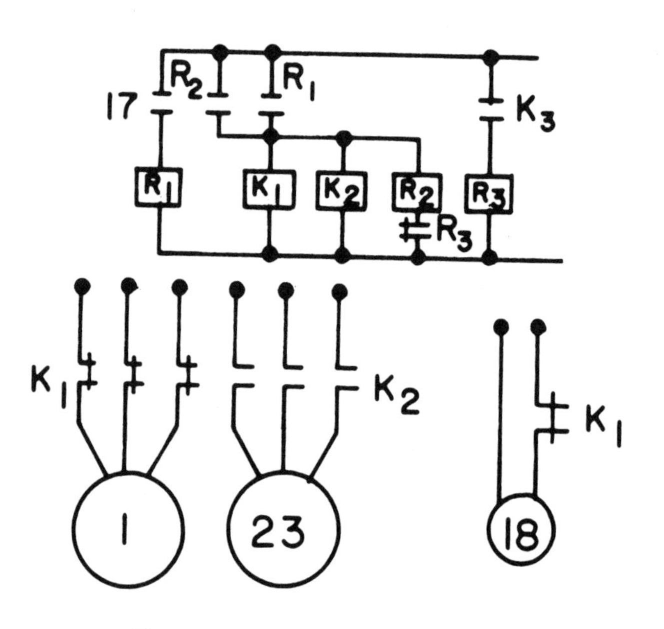

Figure 6-4 Electrical layout of the two motors used to drive the kinematics of the antenna-manufacturing machine.

the spring 19 and prevents friction between the belt 20 and the drum 21). At this time the normally open contacts K_2 are open because both contactors K_1 and K_2 are off. When the desired length of the wire has been obtained and the end of the wire has closed the contact 17, the coil of the relay R_1 is actuated, and its contact R_1 closes the loop and actuates the coils of the contactors K_1 and K_2. As a result the normally closed contacts K_1 must open and the motor 1, together with the coil of the brake 18, is disconnected. The motor 1 stops almost immediately (due to the action of the brake) and the motor 23 begins to rotate, driving the valve 22 and beginning the process of producing an antenna.

Note that after the first bending (cylinders in position *B*) has been carried out, the contact 17 is off, and to prevent the influence of this fact on the action of the circuit, a block relay R_2 is used. Its coil is actuated at the very moment the contact R_1 is closed and the relay R_2 of the parallel contact R_2, which provides the normal continuation of the circuit's work whether the contact R_1 is on or off. When the valve 22 has completed the circle (all operations are fulfilled), a special lever 49 (see Figure 7-3) activates the position switch K_3; this action closes the circuit of the coil of the relay R_3. Its normally closed contact R_3 opens, disconnecting the coil from the relay R_2, which opens the contact R_2 and brings the circuit to the starting position.

The example discussed seems to contradict the statement made earlier that, whenever possible, different, energy sources should not be mixed in one machine. We will now justify our decision. The first part of the kinematic layout is purely mechanical. One can find similar designs in other industrial applications. The second part of the layout is based on the use of pneumatics. The reasons for this choice are:

1. A mechanical solution would be too cumbersome because of the dimensions of the antenna.

2. A mechanical solution is not as flexible as a pneumatic one. If, for instance, one of the dimensions of the antenna has to be changed, the pneumatic solution facilitates such an alteration with only minimal changes in the system; for example, cylinders can be moved if necessary or changes can be made in the diameter of pins with relative ease. Similarly, if we want to produce a part other than an antenna from the same wire on the same machine,

the changes needed will be less for a pneumatic solution than for a mechanical one.

The choice of an electrical layout for synchronizing the work of the motors and brake is obvious.

The design of a kinematic layout often represents a creative process. Sometimes the conventional solution does not fit, and a special solution must be sought. There are also cases where a solution cannot be found in the available printed matter.

The following exercises will help the reader to sharpen his or her skills in this field.

EXERCISES

1. Propose a system that will rotate the lights of a car by an angle ψ, corresponding to the rotation of the steering wheel, through an angle ϕ. *Note:* Small rotations of the steering wheel must not affect the lights (Figure 6E-1).

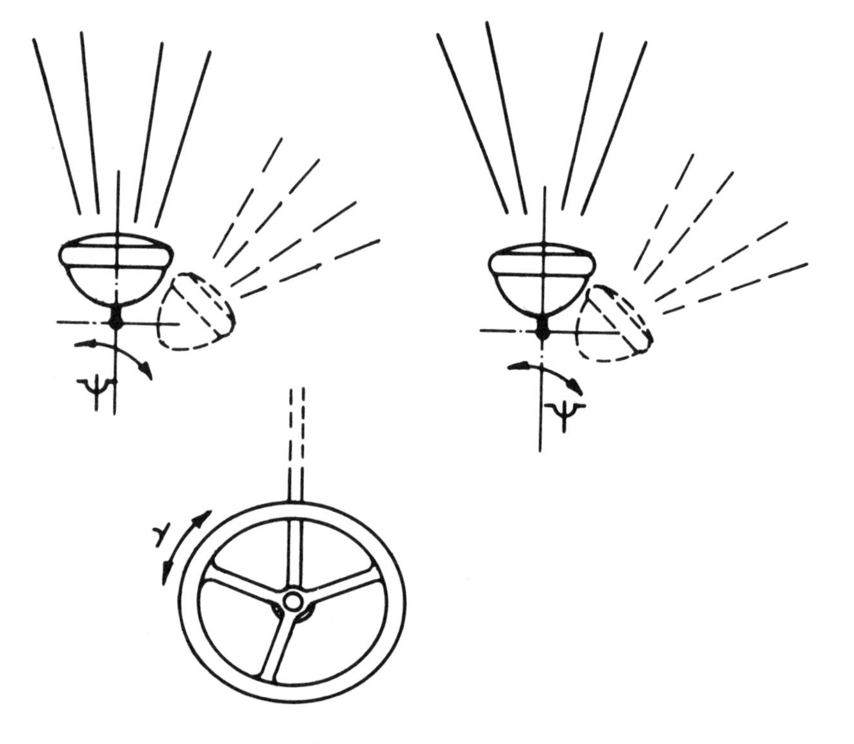

Figure 6E-1 Exercise 1.

2. The diagram shows two handles located as illustrated. Design a device that will permit the movement of only one of the handles when the other one is fixed in the "O" position (Figure 6E-2).

3. The same as Exercise 2 (Figure 6E-3).

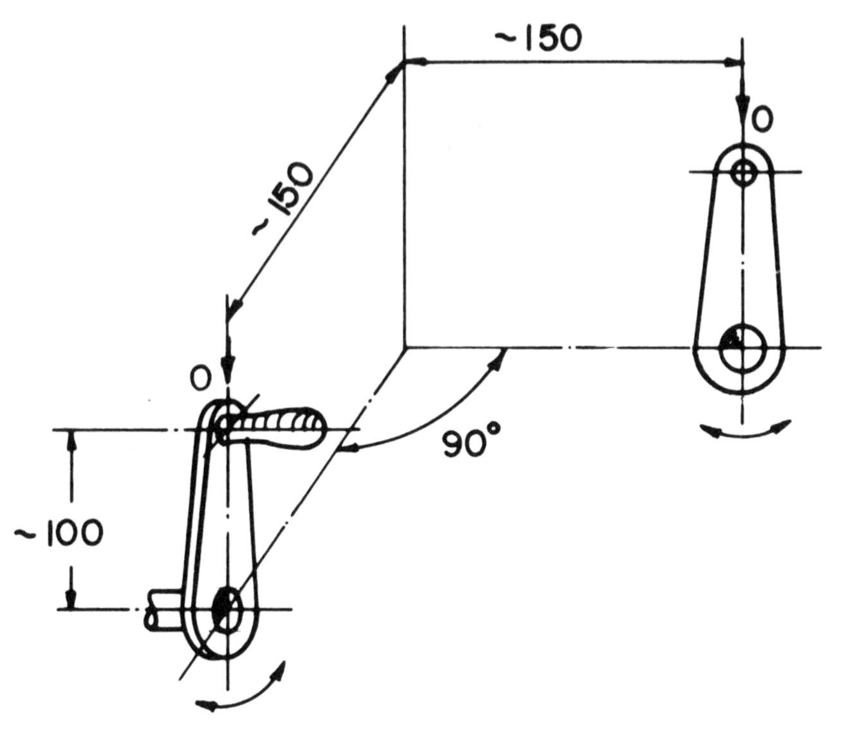

Figure 6E-2 Exercise 2.

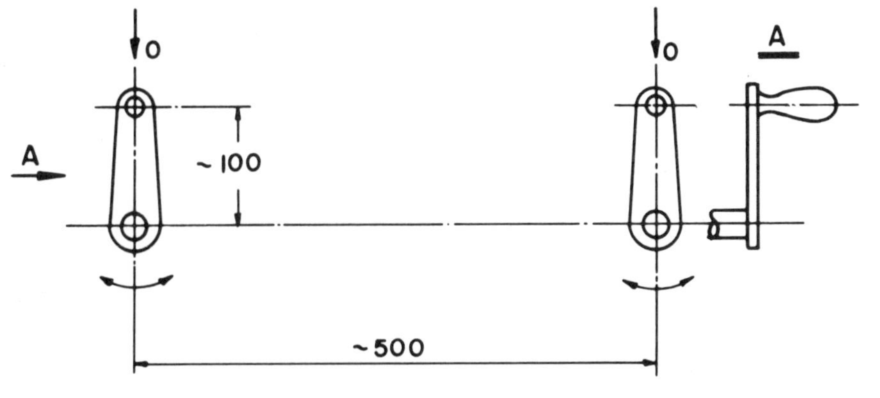

Figure 6E-3 Exercise 3.

4. A number of pushbuttons are given. Find a solution that prevents two or more of the buttons from being pressed simulaneously. Consider the following options:

a. By pressing any button, the previously pressed one is released.

b. A depressed button must be released before another one can be pressed (Figure 6E-4).

5. The layout shows a shaft I rotating at a uniform speed of about 120 r/min and a body II in a plane parallel to the shaft I. Propose a mechanical solution that will enable the movement of the body II (driven by shaft I as shown in Figure 6E-5).

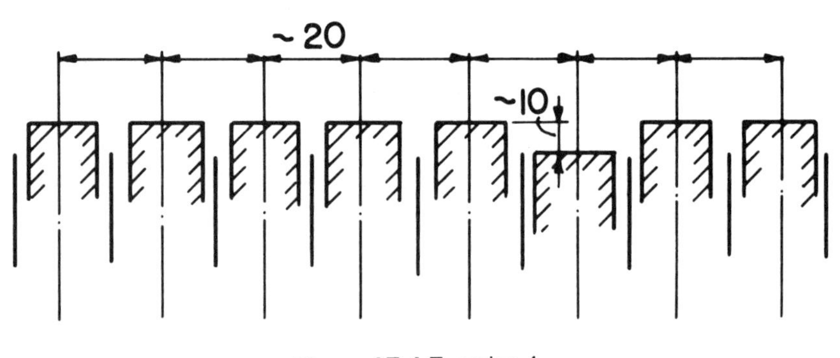

Figure 6E-4 Exercise 4.

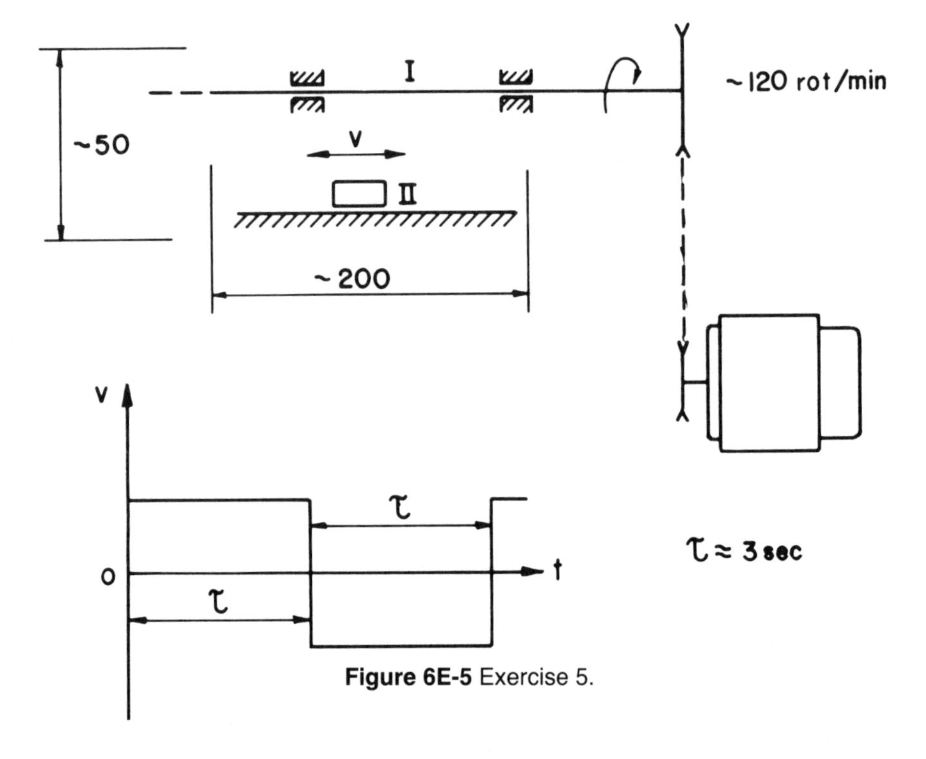

Figure 6E-5 Exercise 5.

6. The diagram shows glass bottles packed in a case. Propose a system for the automated removal of the bottles from the packing case for purposes of cleaning (Figure 6E-6).

7. The diagram shows two pistons I and II housed in a cylinder. Find a solution that will cause piston I to move when $P = P_1$ and piston II to move when $P = P_2$, where $P_2 > P_1$ (Figure 6E-7).

8. The rotating cutter shown in the diagram cuts wire into 20-mm pieces. Propose a solution for feeding the wire and stopping its movement during the cutting process (Figure 6E-8).

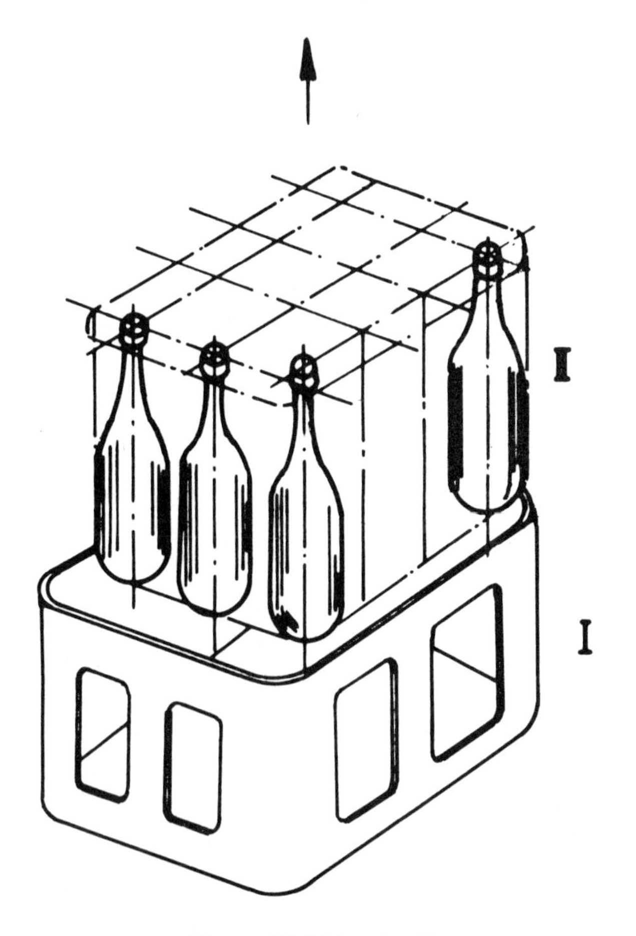

Figure 6E-6 Exercise 6.

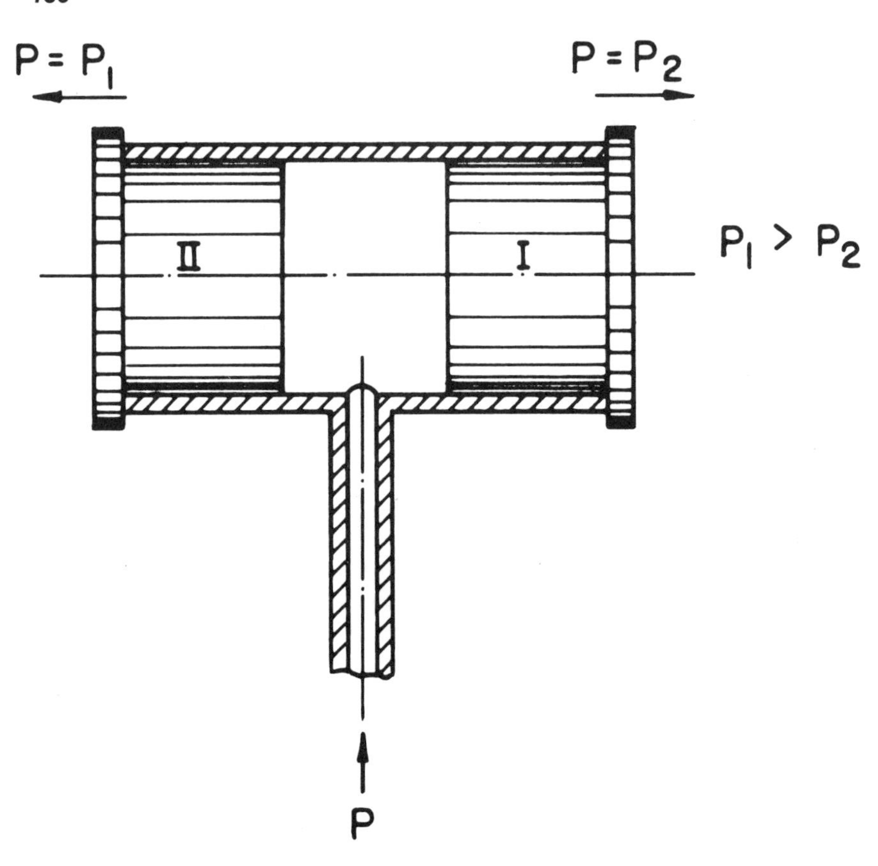

Figure 6E-7 Exercise 7.

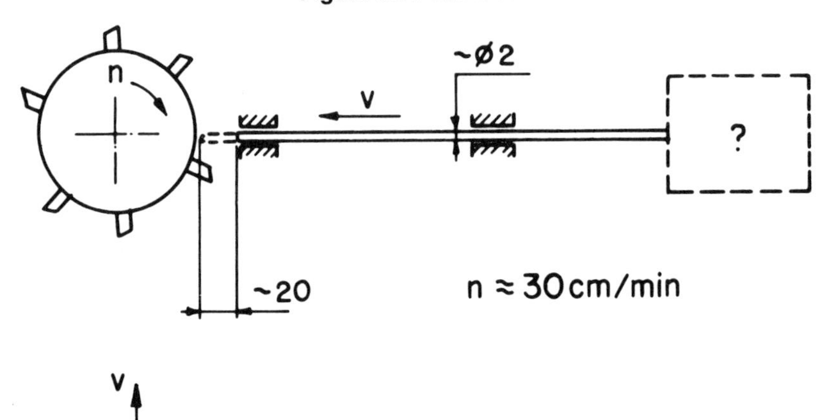

Figure 6E-8 Exercise 8.

9. A coil of wire is given (the diameter of the coil changes as the wire is used up). Propose a system for providing a constant pulling stress F (Figure 6E-9).

10. The layout shows a constantly rotating electric motor and a shaft provided with a tap (for forming an internal screw thread) on its end. Propose a system for facilitating rotation of the shaft with speed n_1 for producing the thread and with speed n_2 for extracting the tap from the threaded part. The axial movement of the shaft is caused by a variable force P (Figure 6E-10).

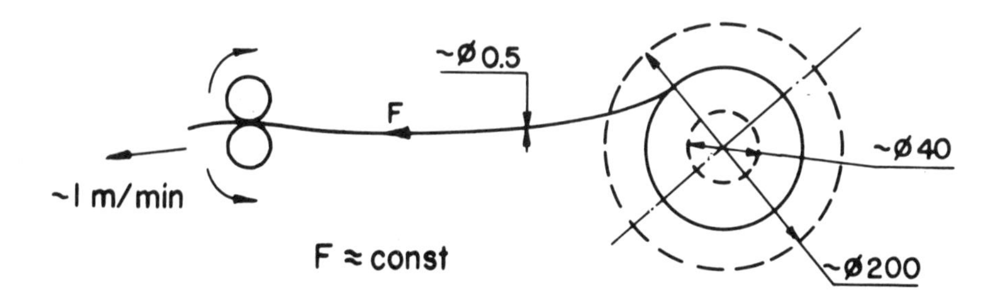

Figure 6E-9 Exercise 9.

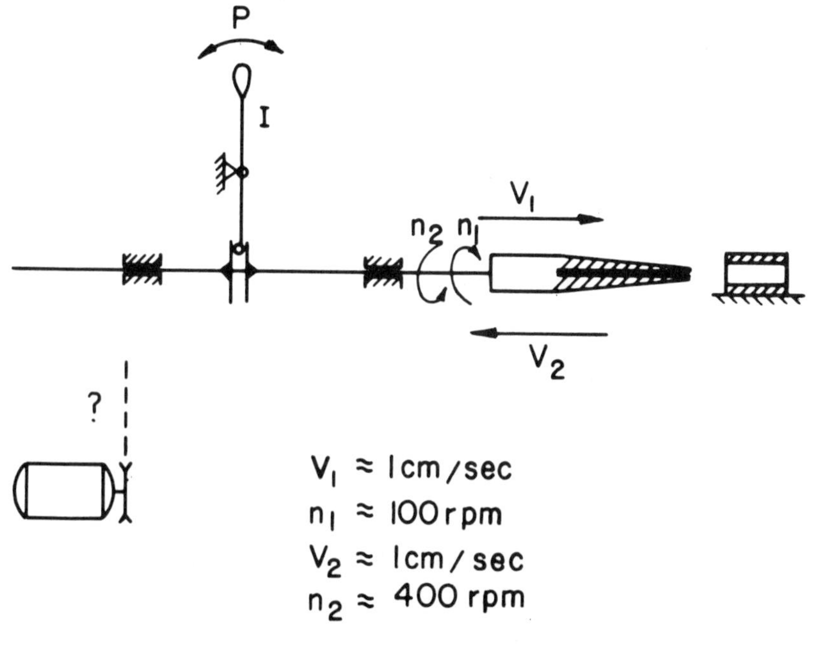

Figure 6E-10 Exercise 10.

11. The diagram shows a punch for cutting certain items out of a moving metal tape. Propose a device for facilitating the movement of the tape in accordance with Figure 6E-11.

12. A set of helicopter blades is shown in the layout. How is it possible that during horizontal flight each blade has an inclination of α_1, α_2, α_3, and α_4 while passing through positions I, II, III, and IV respectively? Take into account that the pilot can change the absolute and relative values of the angles (Figure 6E-12).

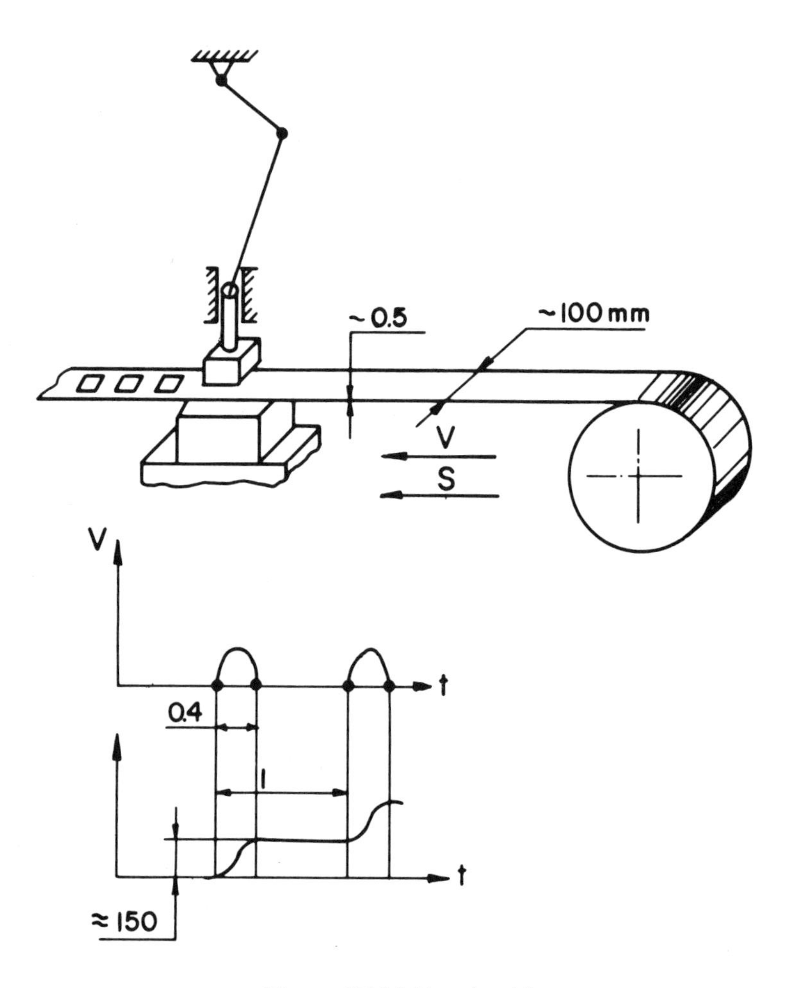

Figure 6E-11 Exercise 11.

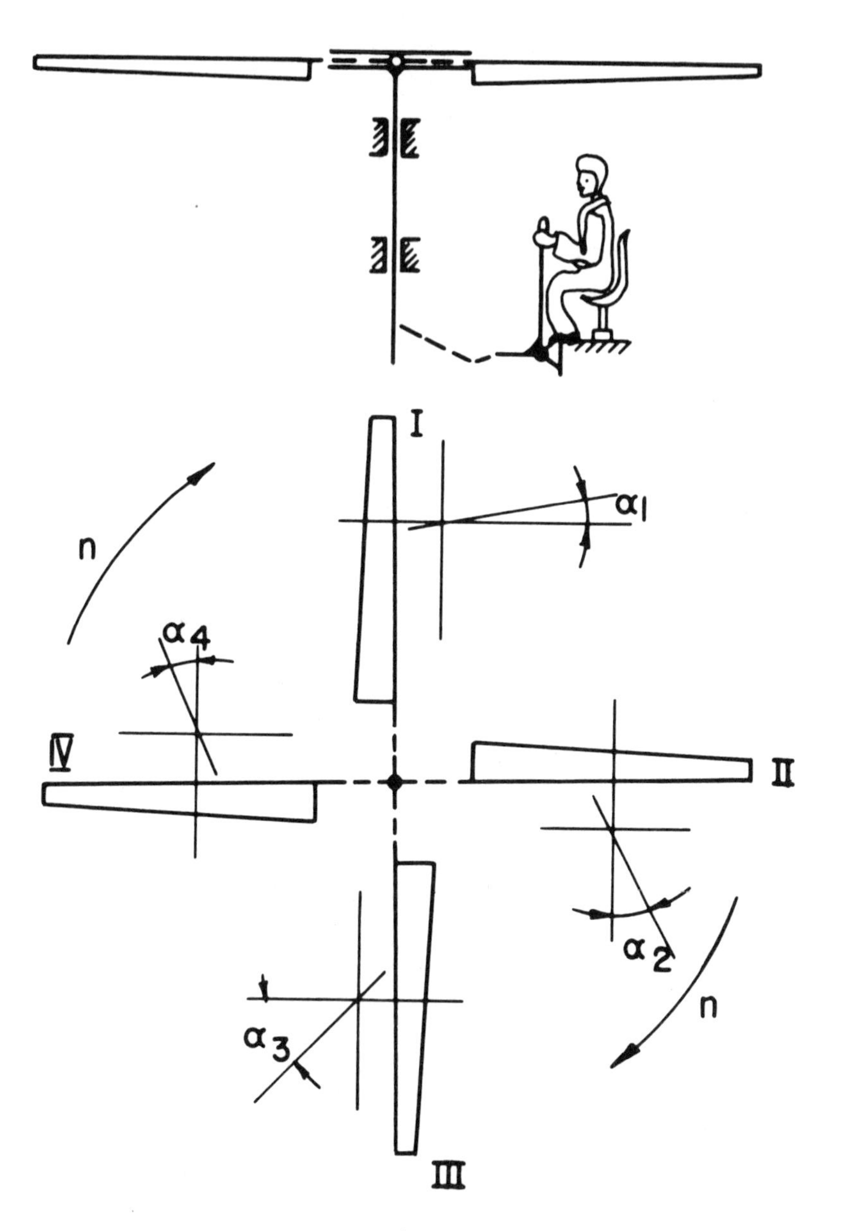

Figure 6E-12. Exercise 12.

13. Two shafts rotating with speeds ω_1 and ω_2 are shown in the figure. How can you arrange for the third shaft to rotate with speed ω_3 in accordance with the given expression (Figure 6E-13)?

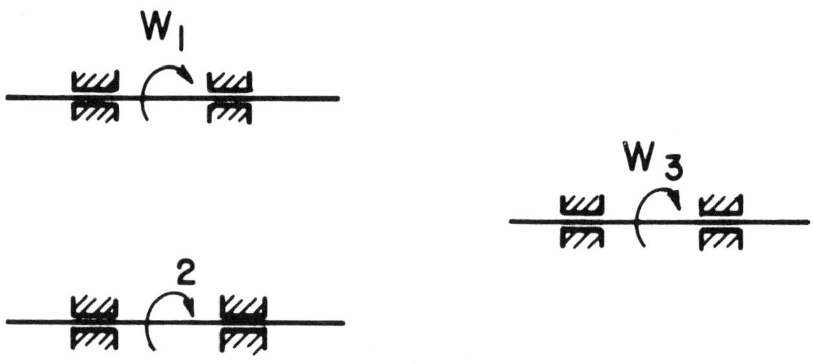

W1

W3

2

$$W_3 = K(W_1 + W_2)$$

$$K = \text{const}$$

Figure 6E-13 Exercise 13.

14. Body I travels as is shown in the diagram with an amplitude of about 150 mm. How can the body II be forced to move in accordance with Figure 6E-14?

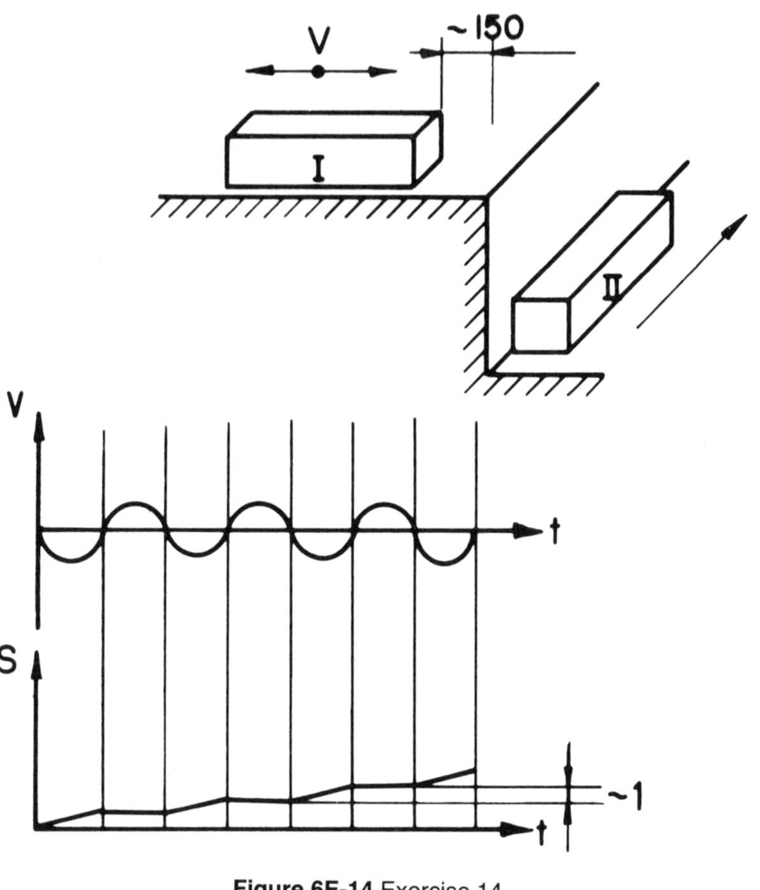

Figure 6E-14 Exercise 14.

15. The diagram shows a drilling device in which a motor constantly drives the drill. Design a system that restricts the torque *M* (if *M* reaches *M*, the drill will have to be disconnected from the drive). (See Figure 6E-15.)

16. Two shafts I and II are shown in the layout. Shaft I has a variable speed ω_1 that vacillates continuously between o and ω_{max}. Design a coupling that connects the shafts when $\omega \geqslant \omega_1^*$ (not before) is reached. In accordance with the diagram when $\omega_1 \geqslant \omega_1^*$, then $\omega_2 = \omega_1$ (Figure 6E-16).

17. Two parallel shafts I and II are given in the diagram. The speed of shaft I is uniform and is equal to ω_1. Design a system that enables shaft II to rotate at speed ω_2 in accordance with the graph (Figure 6E-17).

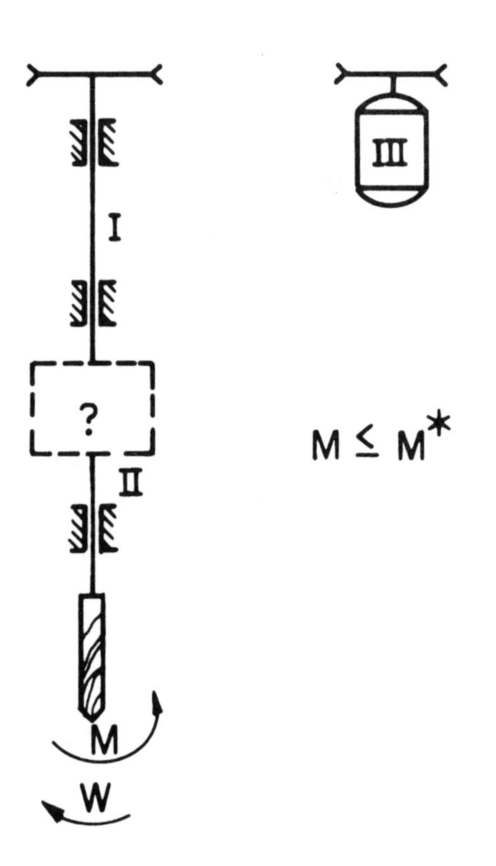

Figure 6E-15 Exercise 15.

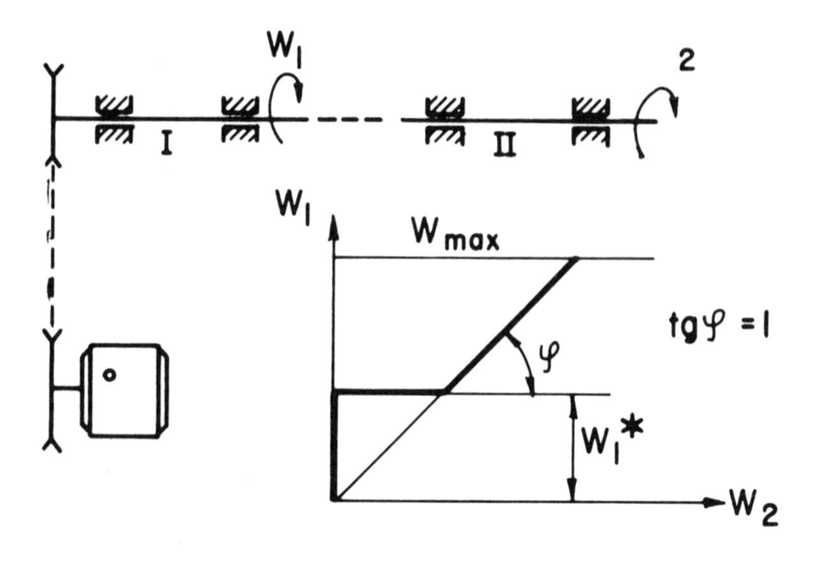

Figure 6E-16 Exercise 16.

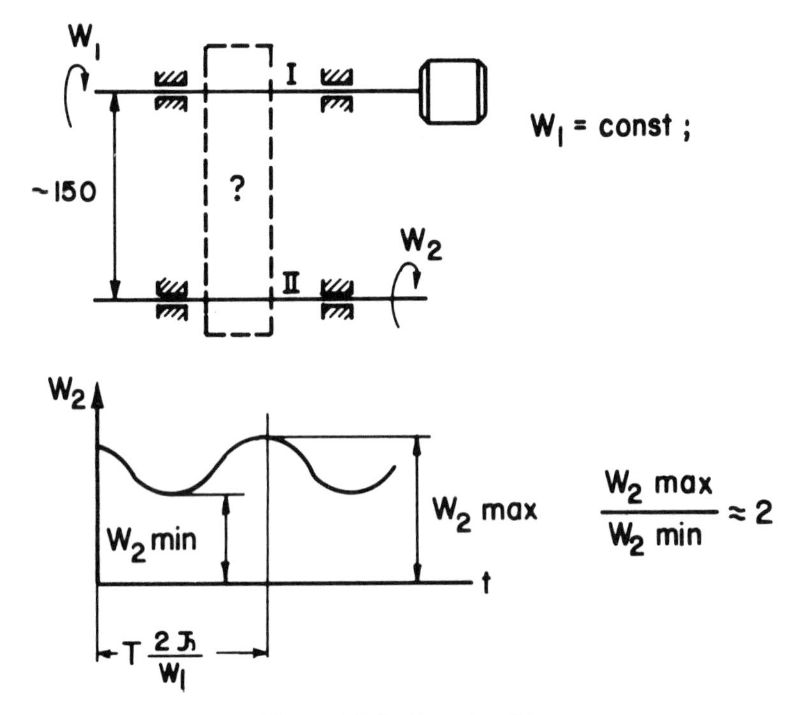

Figure 6E-17 Exercise 17.

18. A pair of railroad car wheels VII and their axes are shown in the diagram. The braking system is actuated by a pneumocylinder VI. Design the transmission from the lever V to the rods I, II, III, and IV, which will force the brakeshoes to be pressed against the wheels during the deceleration process. The transmission must provide a uniform pressing force that is independent of the wear on the shoes (Figure 6E-18).

19. An elevator car is presented in the diagram. Propose a solution for stopping the car immediately if the cable should tear (Figure 6E-19).

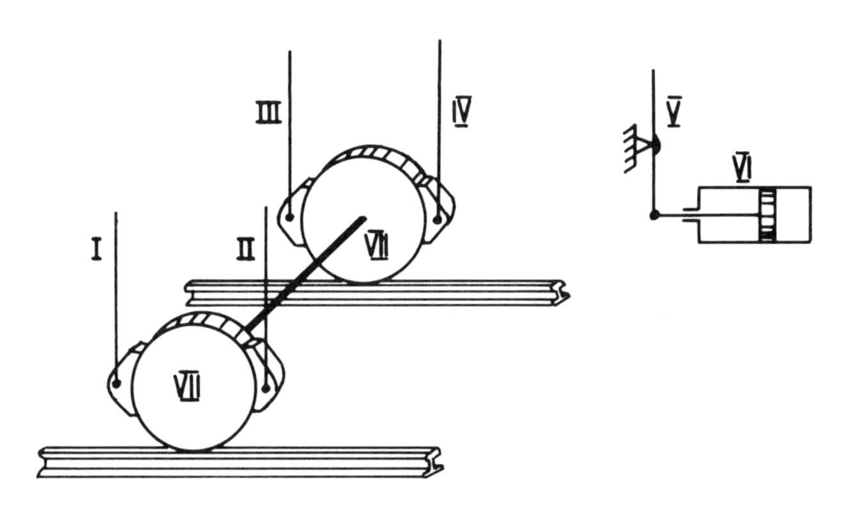

Figure 6E-18 Exercise 18

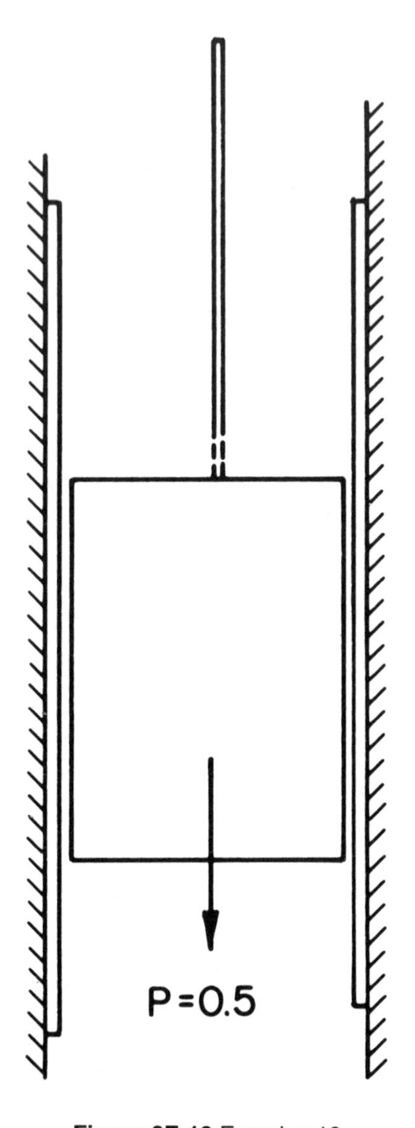

Figure 6E-19 Exercise 19.

20. Two shafts rotating at the same speed $\omega_1 = \omega_2$ are given in the diagram. Design a coupling that allows the shaft free motion around point "o" in the plane of the paper during its rotation (Figure 6E-20).

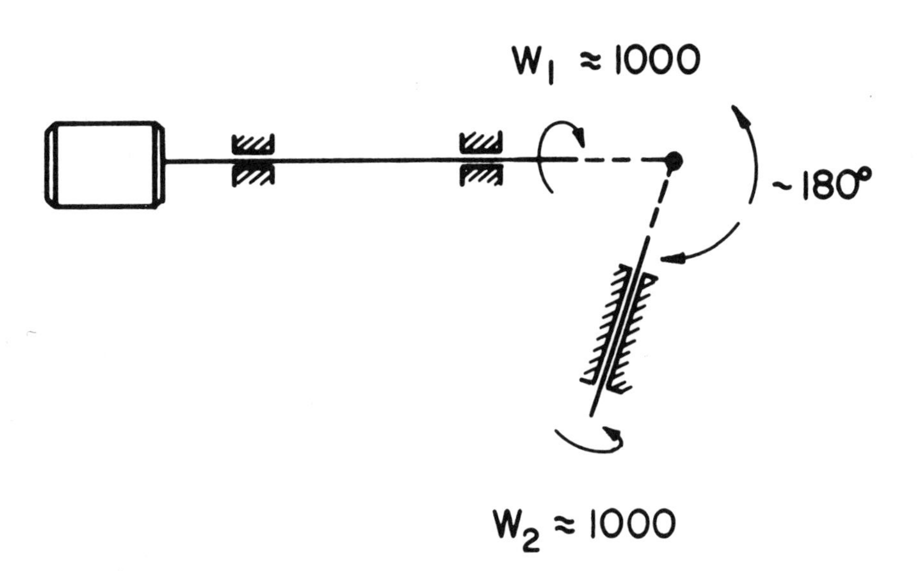

Figure 6E-20 Exercise 20.

21. The diagram shows a propelling device of a boat. An improvement that permits a change in the inclination of the blades during the rotation of the shaft (for reversing or speed changes) is required (Figure 6E-21).

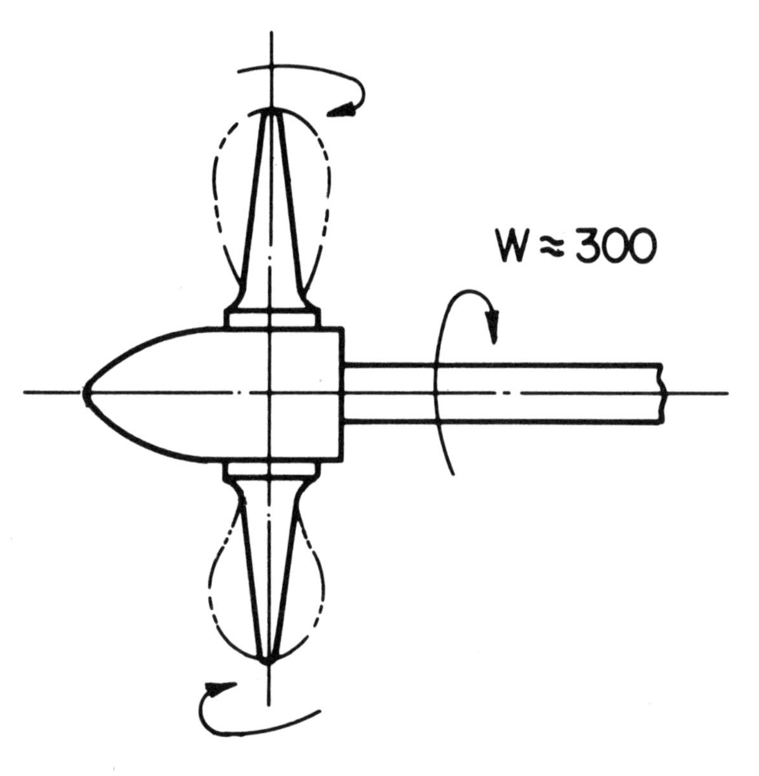

Figure 6E-21 Exercise 21.

22. The layout shows a shaft I rotating at a uniform speed of about 100 r/min and a body II in a plane perpendicular to the shaft I. Propose a mechanical solution that will enable the movement of the body II (driven by shaft I) as shown in Figure 6E-22.

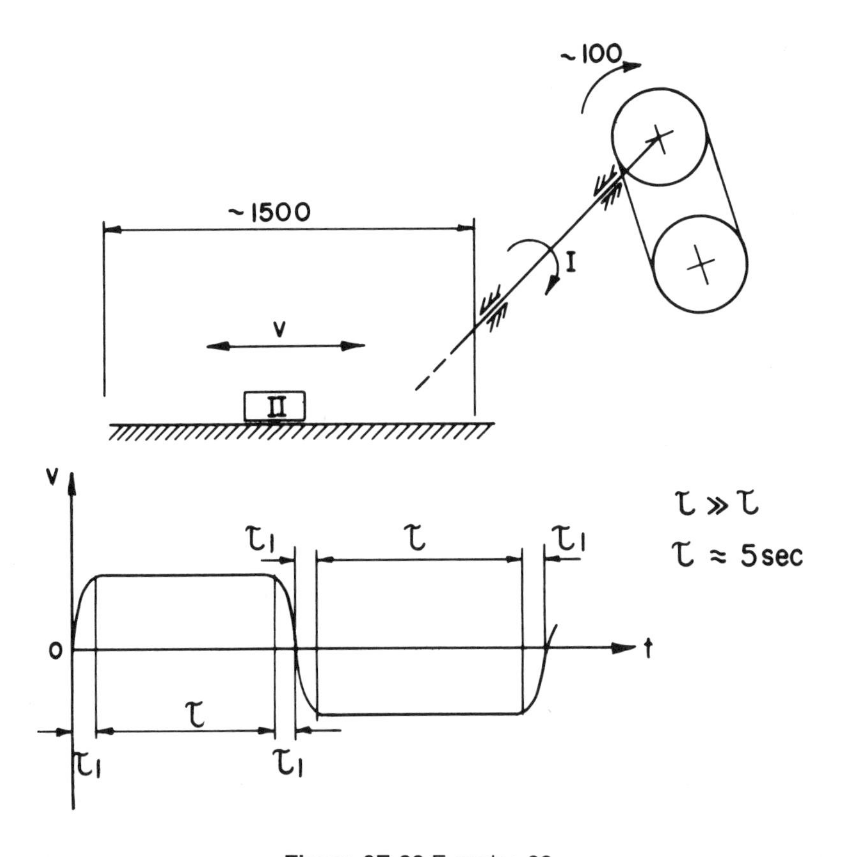

Figure 6E-22 Exercise 22.

Chapter 7

SOME RULES OF DESIGN

When the concepts have been clearly defined and the layouts are ready, the real design begins. The following are some rules the designer must follow in order to create a reliable device. This collection of rules constitutes the "grammar" of design. Most of them do not involve calculations, and so classical engineering books do not deal with them. However, they are based on the experience of generations of engineers, and should be valuable to new engineers to enable them to avoid some of the pitfalls they will encounter in the course of practicing their profession. Here, of course, we can present only a small part of this grammar. The examples are taken from an excellent book written by the Russian engineer Orlov.[6]

The first example illustrates that what sometimes appears to be a negligible element in a particular design may turn out to be of great importance.

Problem: A conical plug located in a housing must provide sealing between the conical surfaces. Which of the four designs shown in Figure 7-1 is the correct one? As we can see, the only differences between the layouts are the relative dimensions of the plugs and the housings. In case 1, after the grinding-in process, the upper diameter *m* of the plug becomes larger than the housing hole (regadless of whether or not the plug material is softer than that of the housing). This fact prevents the plug from moving closer to the housing surface, and, as a result, sealing will not be satisfactory. In cases 2 and 3, after grinding, the lower diameter

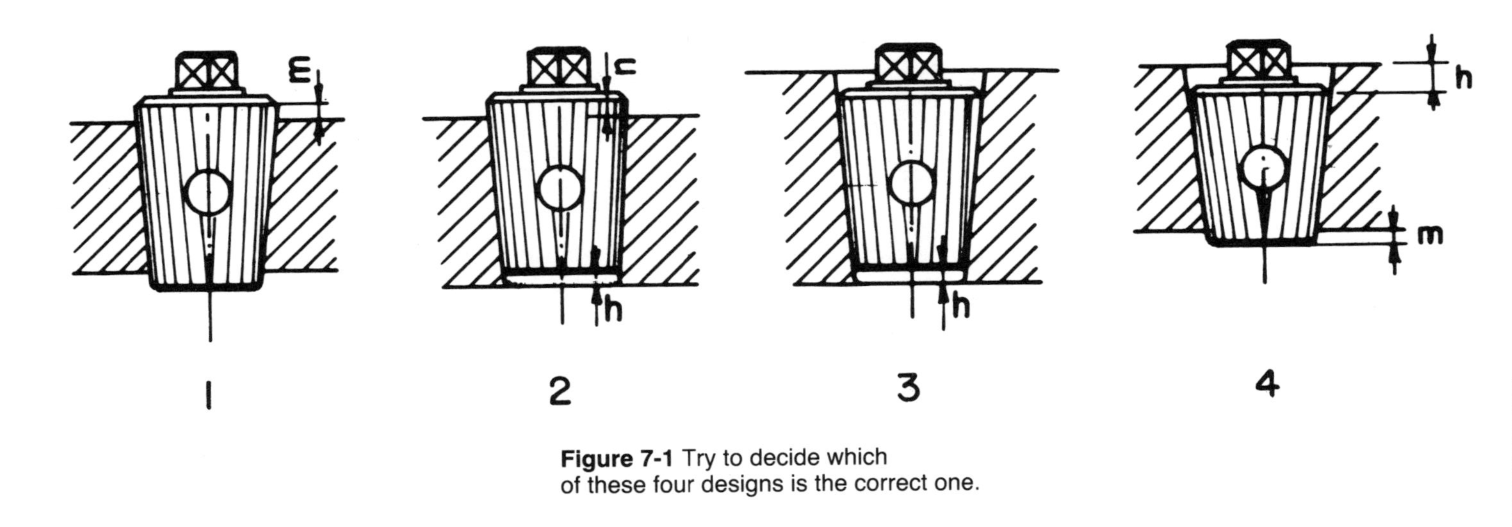

Figure 7-1 Try to decide which of these four designs is the correct one.

h of the housing will prevent the plug from moving deeper, and thus will cause the same problem as in the previous case. The best solution is that shown in case 4: the grinding-in process does not affect the sealing and is independent of the hardnesses of the plug and housing materials—the plug "finds" the best position to close the gap between the conical surfaces.

In light of this example, we can outline one of the most general design rules as follows: *Two adjoining parts should make contact only via one surface each.* Figure 7-2 illustrates this rule. Case 1 is completely incorrect. Case 2, in which backlashes *h* prevent unnecessary contacts, and case 3 offer the correct solutions. Another example of the application of this rule is shown in Figure 7-3. Case 1 is completely wrong, but cases 2 and 3 are satisfactory.

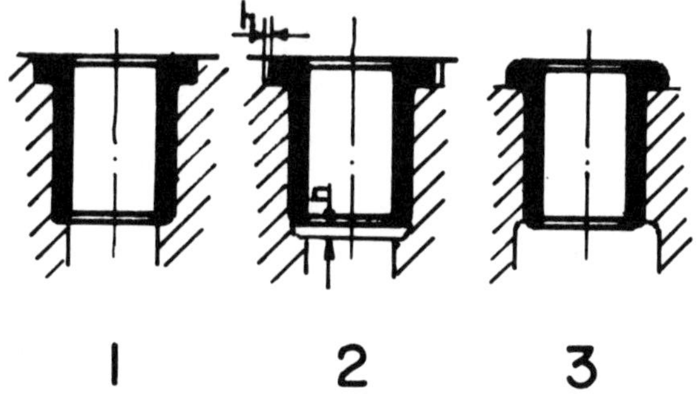

Figure 7-2 Bushing design.

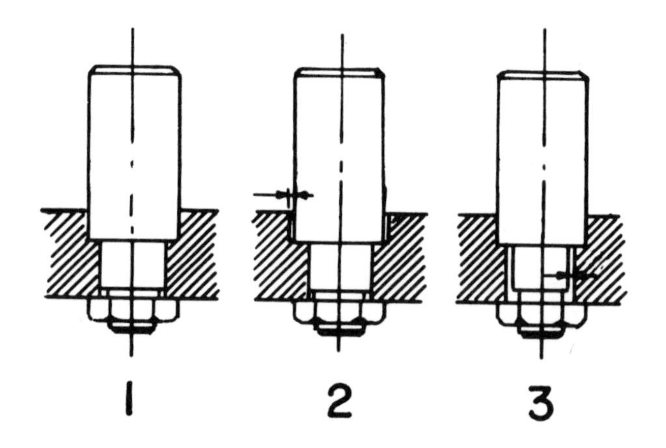

Figure 7-3 Fastening of a pin.

In the coupling shown in Figure 7-4, the connection between the two parts is provided by teeth. Case 1 not only is wrong, but it is harmful because the centering by the cylindrical part of the coupling prevents complete contact between the teeth. On the other hand, in case 2 a backlash S facilitates better contact between the teeth and hence improves transmission of the load.

Another rule states that *a precise movable joint or mechanism has to be released from external forces that can cause intensive wear and disturb normal action of the part*—that is, working surfaces must be protected from the influence of superfluous forces. This rule is illustrated in Figure 7-5 with the example of a conical

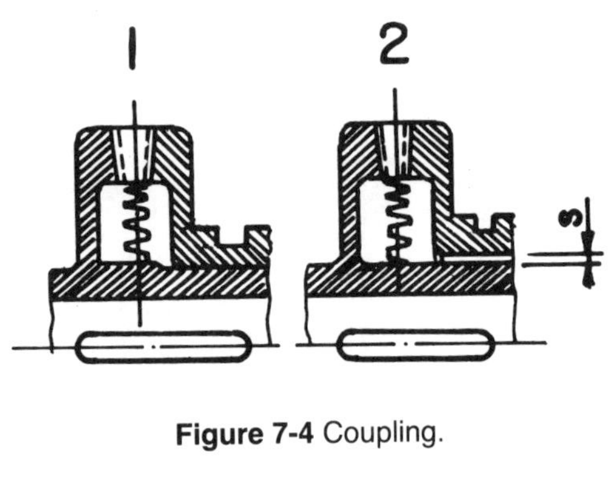

Figure 7-4 Coupling.

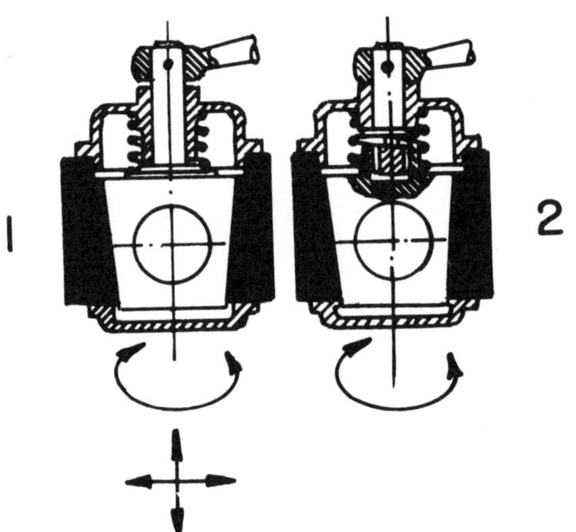

Figure 7-5 A conical faucet with a handle.

plug tap with a handle. In case 1 the turning is precipitated by the conical surfaces. Axial forces applied to the plug by the operator impair the normal action of the tap by causing disturbances in the sealing. The self-centering of the plug is hampered by the drive shaft, which, in turn, is centered by the cover. Consequently the manufacturing process becomes very expensive, since the production and maintenance of the two guiding surfaces require a high degree of accuracy. In the design shown in case 2, the handle and drive shaft are separate from the plug, thus releasing the plug from all forces other than the torque when the handle is rotated.

As another example consider the drive of a valve of an internal combustion engine (Figure 7-6). In the first illustration, the cam pushes the rod of the valve directly. The cam creates a tangential friction force when pushing the valve open. This force causes warping of the valve within the limits of the backlashes. As a result the disk of the valve cannot close the opening promptly, and the exhaust gases flowing through the chink intensify the nonuniform errosion of the valve. In the second case, the valve is designed in such a way that the influence of tangential friction forces on the valve is prevented: only axial, centrally attached forces act on it. This improvement is achieved by the introduction of an intermittent plunger 2. However, this solution entails an increase in the mass of the moving parts. This disadvantage is obviated in the mechanism shown in the third case. Here a lever 3 prevents the influence of tangential friction on the valve.

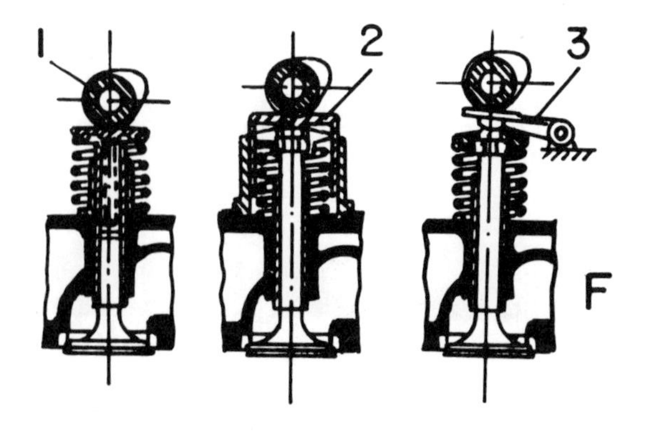

Figure 7-6 Drive of a valve of an internal combustion engine.

A third example is the clamping device shown in Figure 7-7. In case 1, the fixing force acts only at one point on the surface of the object being clamped. In addition the thread of the bolt undergoes bending. The design of case 2 is free of this disadvantage. The ridged surface of the clamping arm finds its position independently of the size of the object. The same applies to the bolt. Also, the clamping process in this case is faster.

The main rule *applicable to the thread is that it must be free of bending and shear stresses.* Let us consider the example presented in Figure 7-8 and compare the six technical solutions. Case 1 is completely unsatisfactory. The maximum bending torque and shear stress caused by the force *P* are located on the threaded

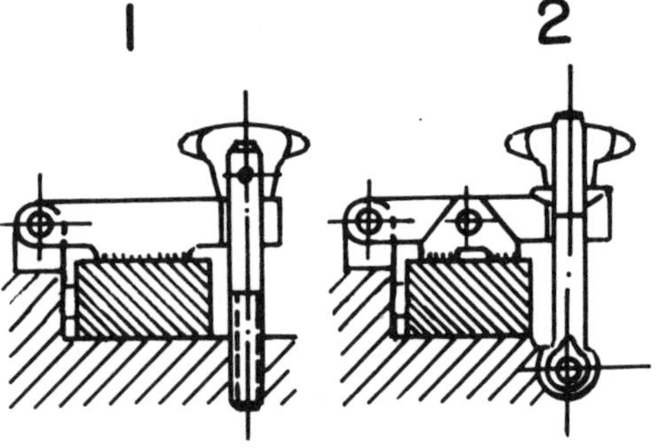

Figure 7-7 Clamping device.

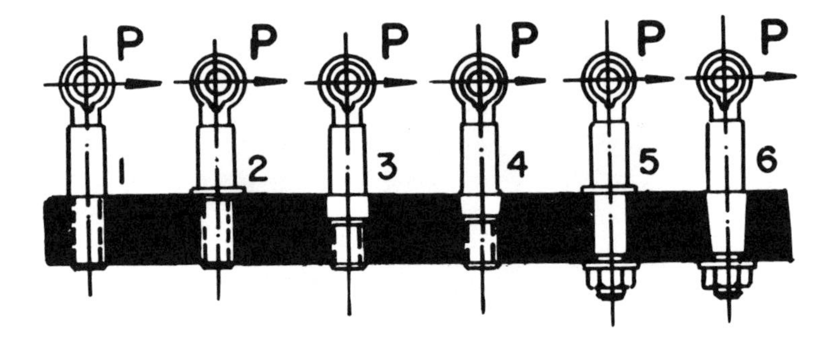

Figure 7-8 Design of a thread.

section of the part. The addition of a collar (case 2) scarcely helps, since the bending deformations are decreased only by the additional friction forces between the collar and the surface of the base. Solutions 3 and 4 are much better. The smooth surface takes the load, thus freeing the threaded section of the part from the bending and shear stresses. The only disadvantage of such a solution is the necessity to provide concentricity of the smooth and threaded sections. Therefore, the thread has to be made with a backlash. The best solutions are offered by cases 5 and 6. Here the thread is loaded only by tension stresses. Bending and shear stresses are taken by the smooth part, which may be conical or cylindrical in shape.

The bending of bolts is often a result of incorrect positioning. The five examples of fixing a support shown in Figure 7-9 will be used to explain this statement. There are two errors in the design of case 1: the lack of an element capable of resisting shear, and the eccentricity of the load due to the force *P* on the bolts, which in turn causes them to bend. In case 2 the base of the support is turned through 90 degrees. Here one bolt is almost not loaded and the other is loaded by the same force that acts centrally. Solutions 3 and 4 are similar, but case 4 obviously is preferable: In case 4 the triangle-like base facilitates the division of the load

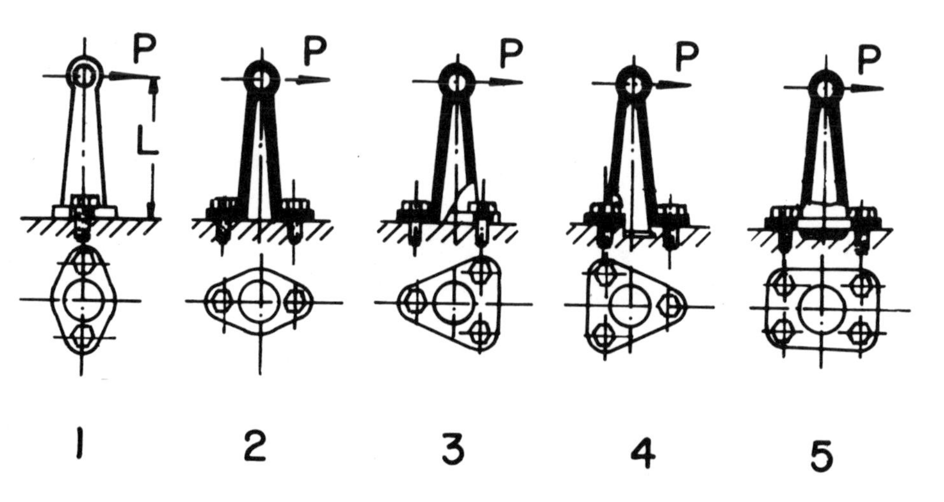

Figure 7-9 Location of bolts.

between two bolts (the disadvantages of case 3 are the same as those for case 2). In addition, in case 4 on the bottom of the support is a lug located in a socket on the base. Case 5 offers the best solution. Here the rectangular base of the support facilitates an increase in the contact area and in the size of the lugs, thus decreasing the load on the bolts and the shear stress.

The designer must avoid, if possible, processes that involve the bending of elements of machine parts, utilizing instead tension, pressing, or shear. For example, it is worth substituting for the lever shown in case 1 of Figure 7-10 the design shown in case 2; the extra rib prevents bending.

Another example is given in Figure 7-11, which illustrates an improvement of a support roller. The best solution is case 3; the worst is case 1.

In the cam follower shown in Figure 7-12, case 2 offers a better solution than case 1. In the latter case, the rod of the follower undergoes considerable bending, which worsens the action conditions of the mechanism. The follower can be jammed in its guide rails.

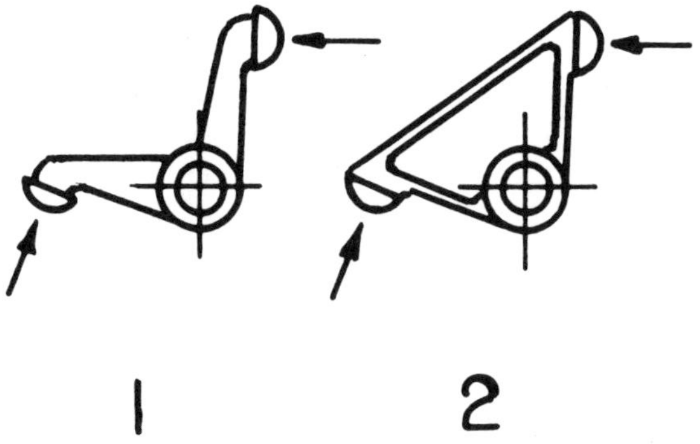

1 2

Figure 7-10 Design of a lever.

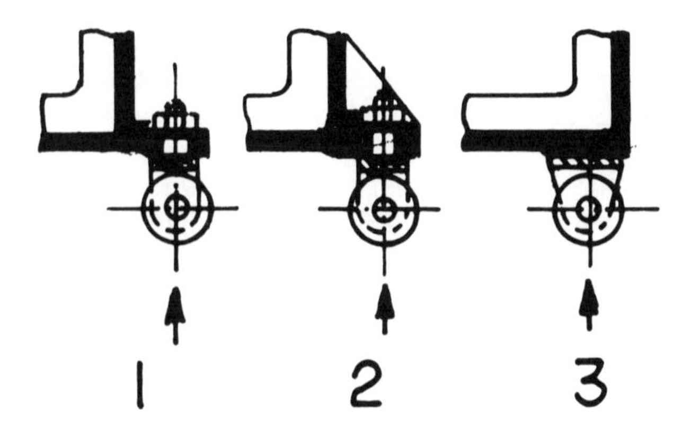

Figure 7-11 Design of a support.

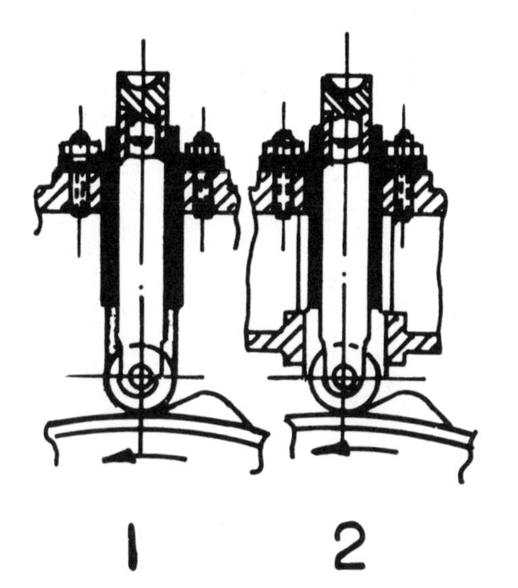

Figure 7-12 Design of a cam follower.

The fast rotating wheel shown in Figure 7-13 must be designed in accordance with diagram 3. In cases 1 and 2, the centrifugal tension forces due to rotation will cause spatial bending of the disk.

The designer must avoid axial fixation of any part at two points along its length, especiallyif the points are far apart. Such a situation is typical for bearing assemblies. Figure 7-14 shows a comparison of the correct (*2*) and incorrect (*1*) designs. The former allows the left bearing to find its place in accordance with the thermal expansion of the housing and the shaft. In addition, errors that may occur during the manufacturing process will not influence the normal action of this assembly.

A part should be centered on a base by a surface with the smallest possible diameter to increase the accuracy of centering and decrease the influence of the temperature changes. Figures

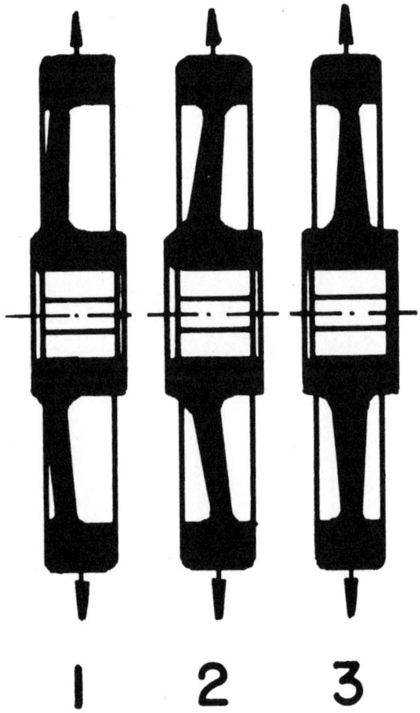

1 2 3

Figure 7-13 Design of a flywheel.

7-15 and 7-16 illustrate this statement. In both cases design 3 is better than 2, with the latter preferable to design 1. For instance, if the support presented in Figure 7-16 is centered on its larger diameter, say about 200 mm, the backlash should be about 0.12 mm: in the optimal case, the backlash can be made three times less, and for the same fit it will be about 0.037 mm. This example should serve to clarify the nature of the rule under consideration.

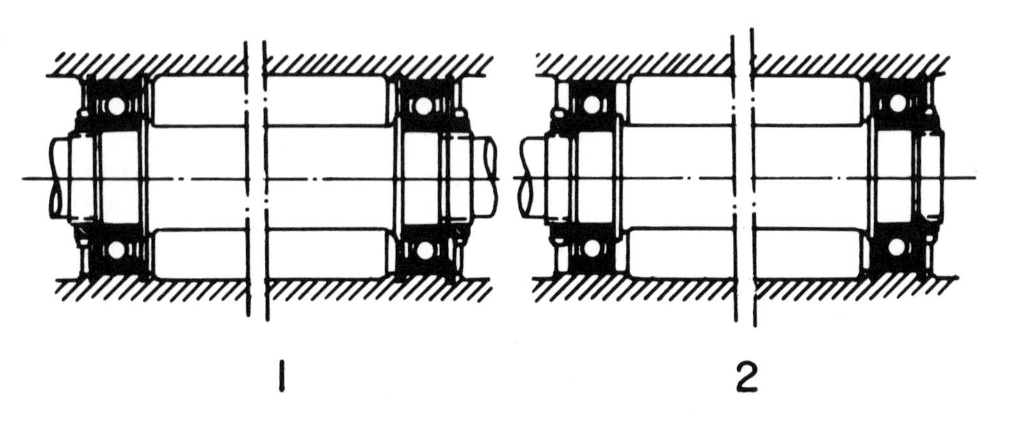

Figure 7-14 Design of a bearing assembly.

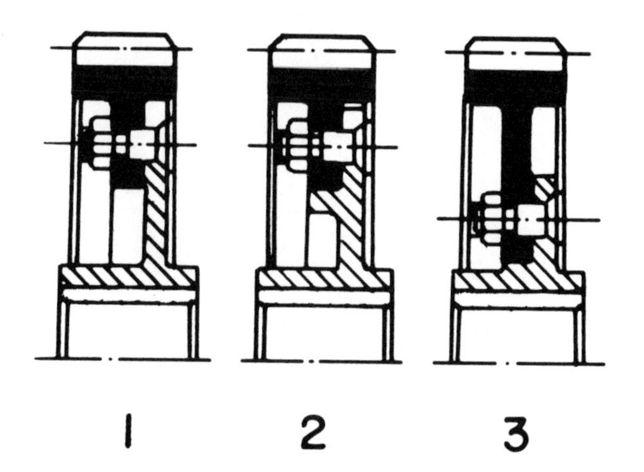

Figure 7-15 Gear-wheel design.

When designing an assembly on the principle of shrink, the designer must remember that the assembly will be under load, even before the working load is attached. It is unacceptable to fit a part on two cylindrical surfaces of equal diameter, as shown in Figure 7-17, case 1. The sizes of the surfaces must be different, as in case 2. In addition, the axial dimensions of the part should be chosen in such a way that the narrower end is fitted into its hole before the wider end enters its hole, as shown in case 3.

Similarly, it is incorrect to design an assembly corresponding to Figure 7-18, case 1. To save the cost of precision processing, the designer has divided the contact surface between two cylinders. The error lies in the fact that the diameters of these cylinders are equal. When the lower cylinder passes through the hole, it will be damaged and will not be able to function properly. Cases 2 and 3 not only offer a better solution, but they also provide higher accuracy.

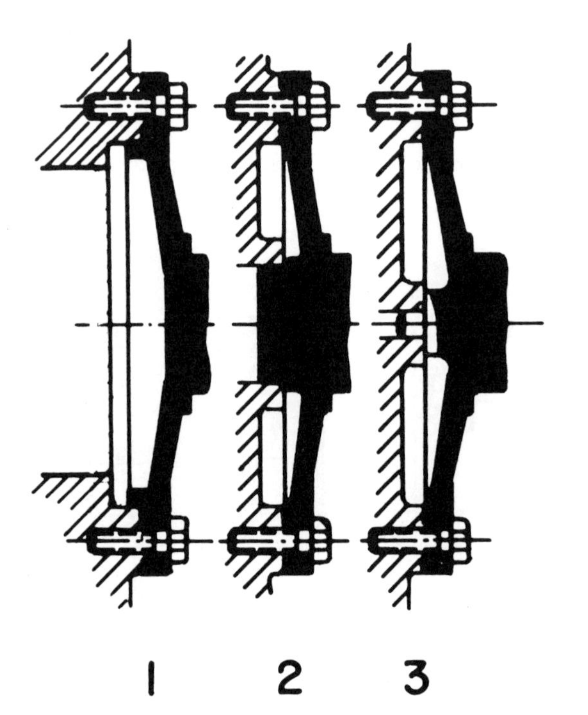

Figure 7-16 Design of a cylindrical support.

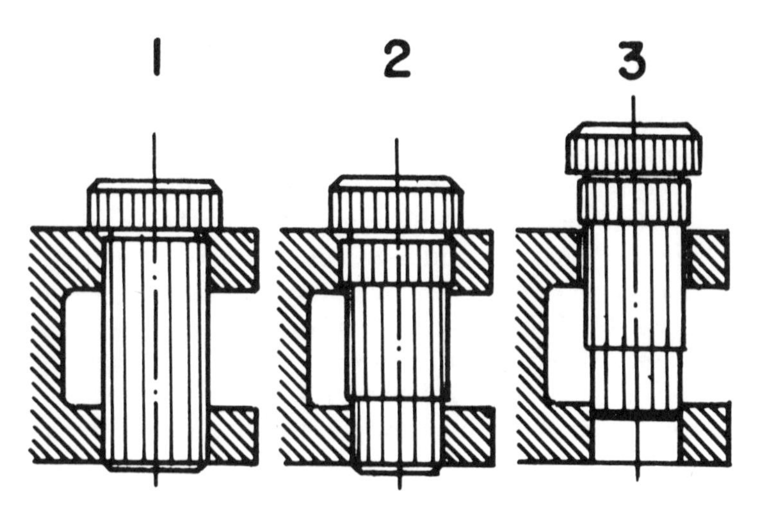

Figure 7-17 Assembly of a pin.

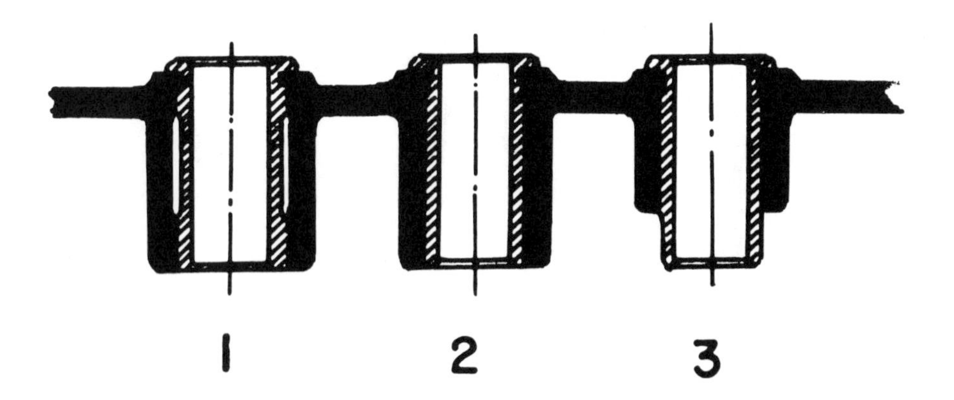

Figure 7-18 Assembly of a sleeve.

Thermal stresses cause serious problems in the construction of machines, especially parts that are subjected to changes in temperature. The following example will help us to understand this point. Figure 7-19 shows the cooling jacket of the cylinder of an engine. The temperature differences between the internal and external surfaces, and hence differences in thermal expansion, give rise to stress. A possible solution lies in the use of a material with a very small expansion coefficient, but in most cases this solution is not acceptable because of other restrictions. The solution offered by case 1 is incorrect because of the straight shape of the walls. Here the stresses reach high values. The shape of the external wall in case 2 facilitates an increase in the elasticity of the jacket and a consequent decrease of the stress.

Some "grammar" *rules relating to cantilever shafts can be explained with the help of* Figure 7-20. It is easy to see that in case 1 the reaction forces in the bearings are significantly different, which results in the bearings having different lifetimes. (The use of different bearings complicates the device.) Solution 2 improves the relationship between the forces. In principle they can be made

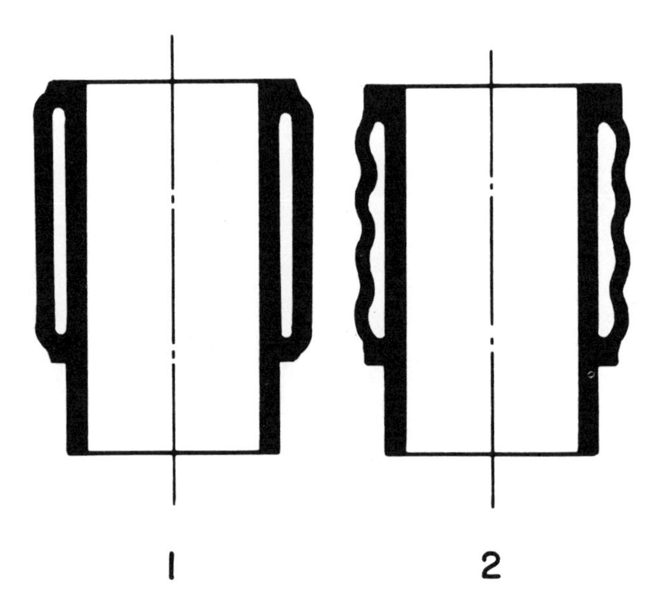

Figure 7-19 Cooling jacket.

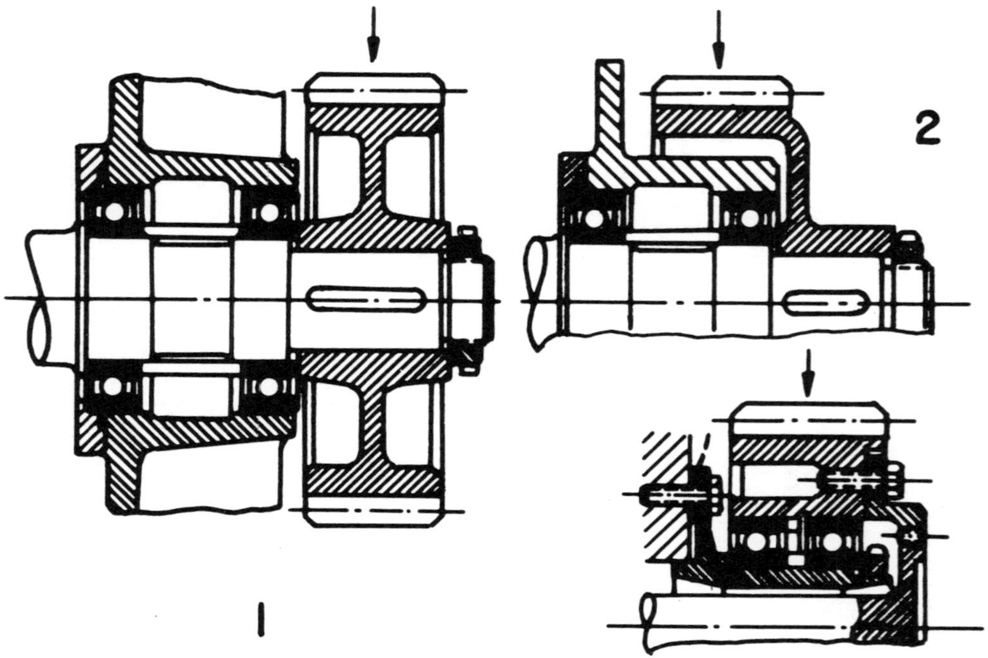

Figure 7-20 Design of a cantilever shaft.

completely equal. The third case eliminates the cantilever. The loads on the bearings are identical and the bending stresses in the shaft are eliminated. The shaft is loaded only by torque. The disadvantage of this solution is the shorter life of the bearings as a result of the rotation of the external rings. (Try to answer the question: Why is this kind of rotation worse?)

The inversion method is a very useful tool in design, and a "grammar" rule based on it may be formulated as follows: *To improve the design, try to invert the chosen solution.* Sometimes this rule permits the designer to achieve the required effect in a less expensive way. The pros and cons of each possibility should be weighed before making a final decision.

This rule is illustrated in Figure 7-21. In case 1 the rod is driven by a forklike lever that pushes or pulls the pin fastened to the rod. The inverse design 2 differs in the location of the pin. Here the pin is fastened to the lever and the fork to the rod. Design 2 is better from the point of view of the distribution of the forces. On the other hand, the advantage of the first design is that it has smaller dimensions.

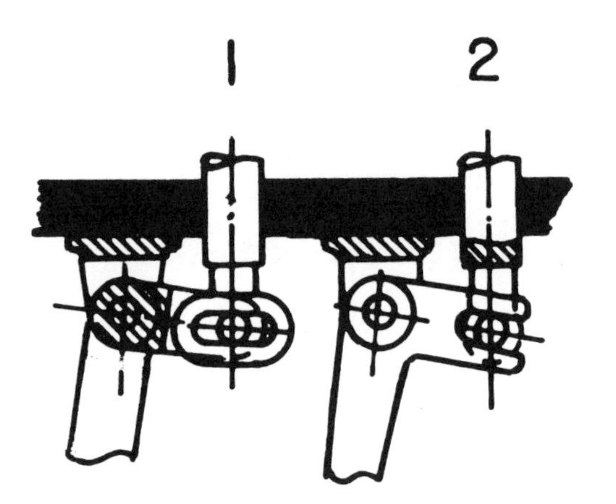

Figure 7-21 Design of a forklike lever.

In Figure 7-22 cases 1 and 2 show two variations of a pipe joint. In the first design, there is an external screwnut, and in the second design, an internal one. The first design has larger diametric dimensions but smaller axial ones. The second mode is smaller in diameter but greater in length.

Figure 7-23 shows two possibilities for fastening a turbine blade. The first provides better conditions for manufacturing the wheel and the second for production of the blade roots.

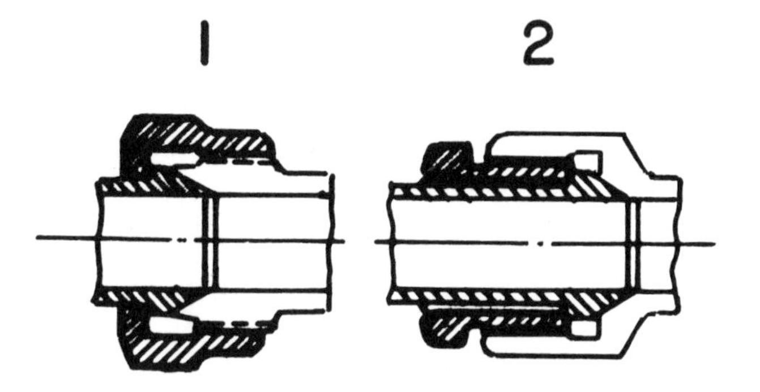

Figure 7-22 Design of pipe joints.

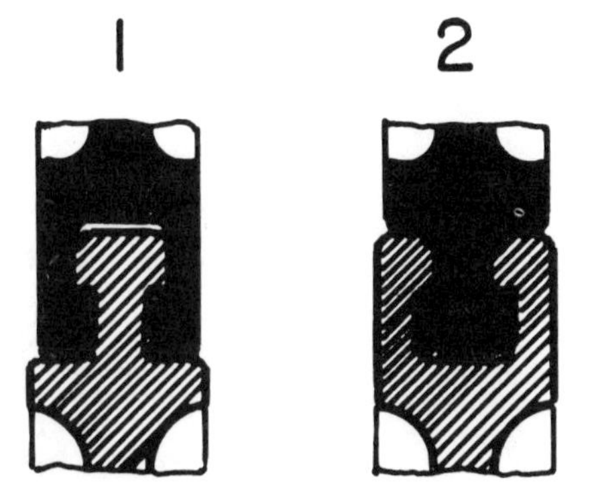

Figure 7-23 Fastening of a turbine blade.

Figure 7-24 illustrates two solutions for connecting a rod to a forkpiece. In the first solution, the connecting rod has one bearing; in the forkpiece, the pin is fixed. The dimensions can be made smaller but lubrication then becomes a problem. In the second case, the rod is fixed on the pin and the bearings are placed in the fork piece. The dimensions are increased, and in general lubrication is facilitated.

The hydraulic cylinders presented in Figure 7-25 differ in that

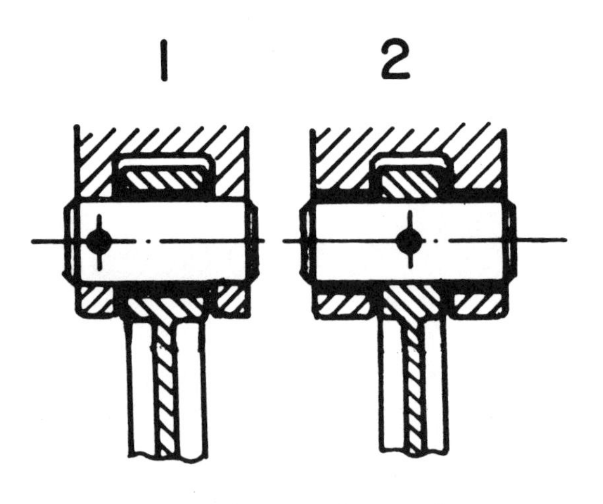

Figure 7-24 Means of connecting a rod to a forkpiece.

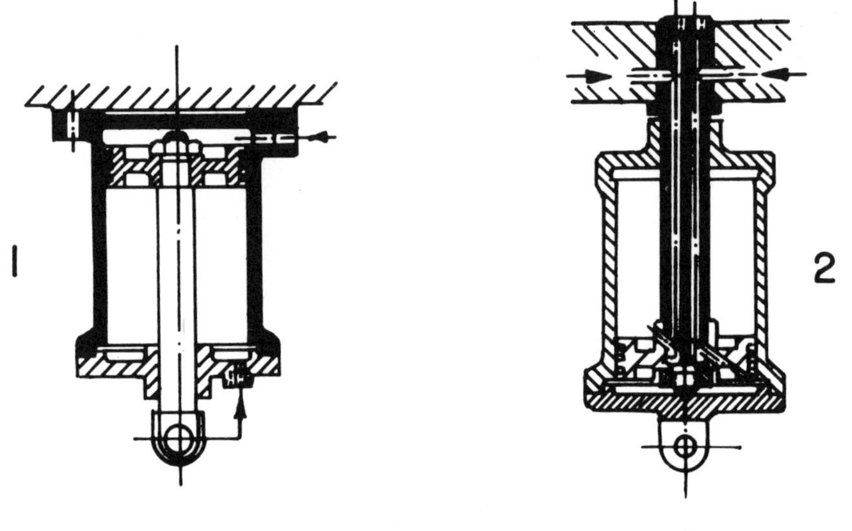

Figure 7-25 "Direct" and "inverse" hydraulic cylinder.

in case 2 when pressure is applied, the cylinder moves while the piston rod is stationary. In the first, more usual, case, the principle is the inverse: when pressure is applied, the piston and the piston rod move, while the cylinder is fixed. Under certain conditions this inversion is very useful.

EXERCISES

Which of the following proposed designs is the best? Try to explain your decision.

1. A bolt joint (Figure 7E-1).
2. A cylindrical support (Figure 73-2).
3. Securing of a housing (Figure 7E-3).

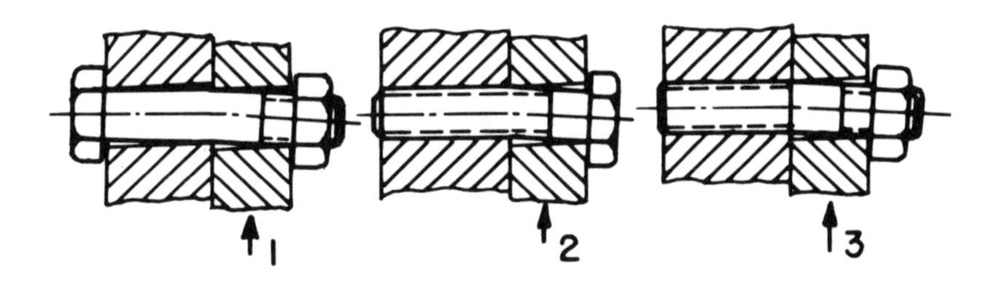

Figure 7E-1 Execise 1.

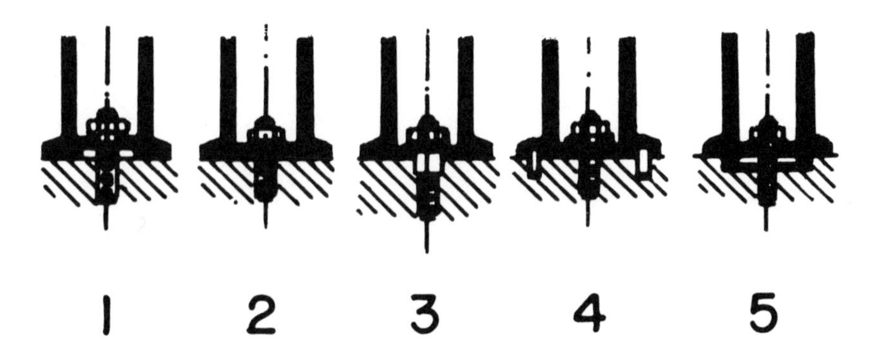

Figure 7E-2 Exercise 2.

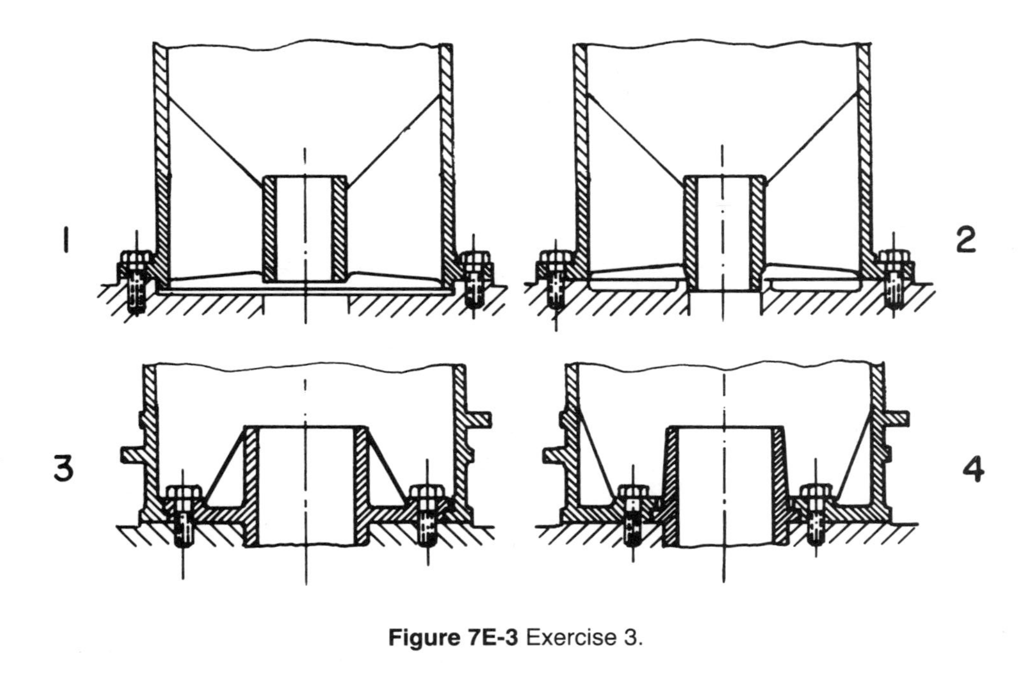

Figure 7E-3 Exercise 3.

4. A housing subjected to different temperature conditions (Figure 7E-4).

5. A welded lug (Figure 7E-5).

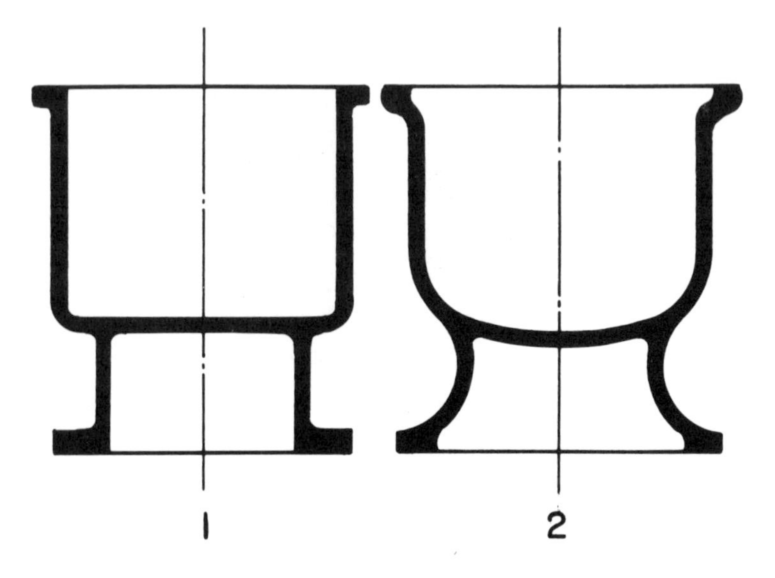

Figure 7E-4 Exercise 4.

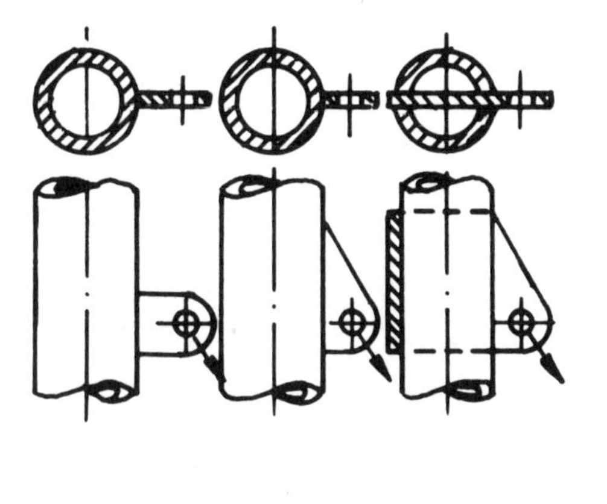

Figure 7E-5 Exercise 5.

6. A joint element (Figure 7E-6).
7. A two-step plunger of a compressor (Figure 7E-7).

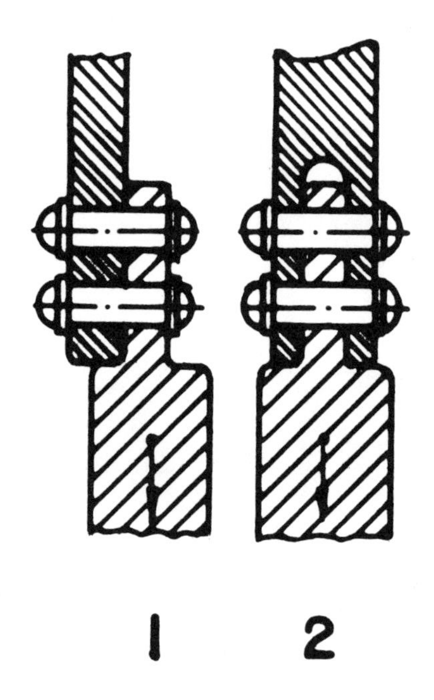

Figure 7E-6 Exercise 6.

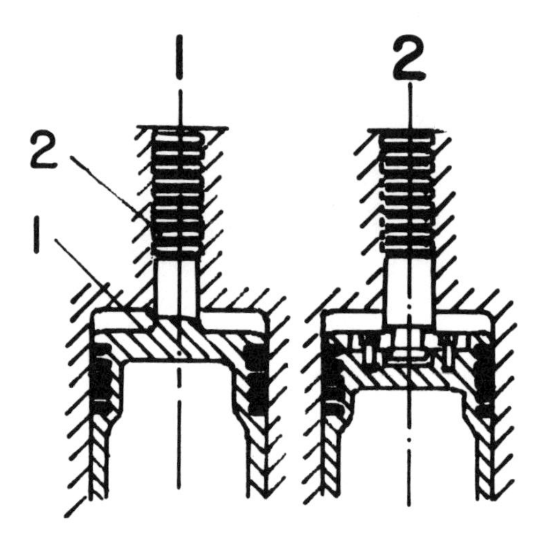

Figure 7E-7 Exercise 7.

8. A cover closing either of two pipes (Figure 7E-8).
9. The roller assembly of a cam follower (Figure 7E-9).

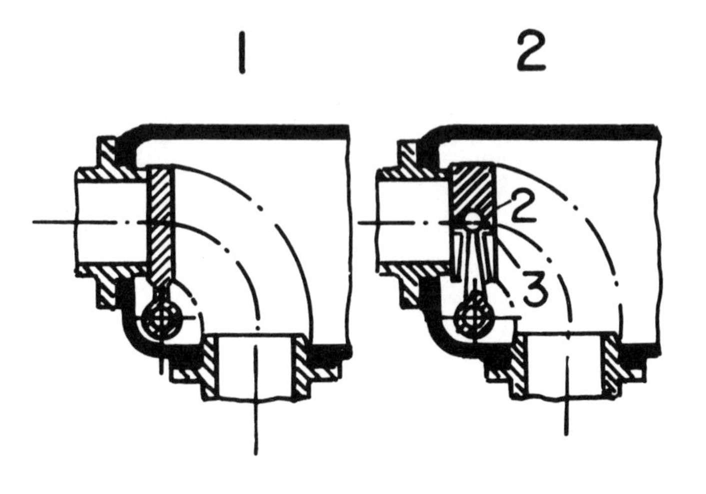

Figure 7E-8 Exercise 8.

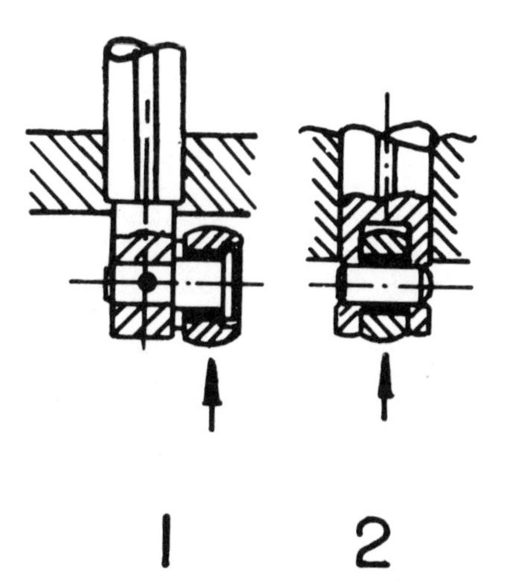

Figure 7E-9 Exercise 9.

10. A thrust-bearing support (Figure 7E-10).
11. A conveyer link joint (Figure 7E-11).

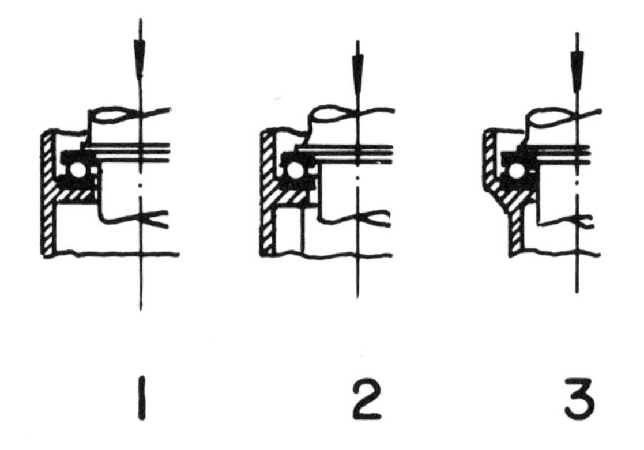

Figure 7E-10 Exercise 10.

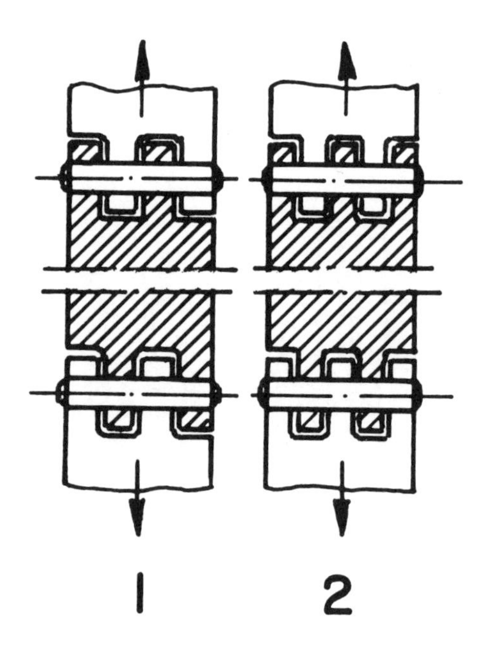

Figure 7E-11 Exercise 11.

Chapter **8**

COMPUTERS IN TECHNICAL CREATIVITY

No discussion of any technical process is complete these days without considering the possible role of the computer, that ubiquitous superservant that promises to take all the drudgery out of life—from planning our meals, and cooking them, to managing our finances. Thus our question here is: How can the creative process be computerized, if at all? The key word, of course, is "creative." It seems we know more about why it is not feasible to computerize creativity than about how computers can be applied in creative design.

The natural conclusion to this book is a consideration of the role of computers in creative design. This chapter thus is devoted to some ideas about the nature of creativity in general, and of technical creativity in particular, and about how creativity can be computerized—if at all.

Creativity produces information and knowledge, and whoever controls information and know-how will inherit the productive world. Computerization of this process of knowledge production could contribute much to science and technology. The process of creating a new product, machine, or technical solution can be described as a "game" with nature, a game in which nature is essentially an indifferent partner. Thus we confront the unknown

without a definite apponent. Richard Bellman[1] has expressed this idea differently: "The assumption that the universe is completely hostile is not a feasible one, since it is too expensive."

After beginning our in-depth discussion of the subject with a brief description of the state of the art of computer-aided activities in engineering, we compare the known applications of computerization with the less-well-investigated field of the role of computers in technical creativity.

CAD, CAG, AND CAM—STATE OF THE ART

In modern engineering we are familiar with a number of computer-aided processes:

computer-aided design (CAD);
computer-aided graphics (CAG); and
computer-aided manufacturing (CAM).

CAD can be applied in industry and technology for the design of machines, apparatus, circuits (pneumatic, hydraulic, mechanical, electronic, etc.), tools, structures, and systems. It is particularly useful in the design of gear transmissions and cam mechanisms, turbines and missiles, engines, and manufacturing processes and plants. In machine design the following specific domains of computerization are worthy of mention.

1. CAD can be applied to the investigation of the kinematics of mechanisms and machines, which includes the study of the motion of the links of the mechanism. For example, it is often very important to define the displacement of some specific point or element as a function of time (or of some other displacement), the trajectory of a specific point, the velocities of certain links, or, of course, the accelerations of links, which are the sources of dynamic loads, vibrations, and noises.

2. The forces acting between and within links constitute another important factor that must, and can, be elaborated by means of a computer. The importance of this application of CAD cannot be sufficiently emphasized.

3. Computerization can also be used in the design of machine parts, including the determination of their main dimensions, the choice of materials (as well as special treatments), and, often, the mode of manufacture.

To help us understand this brief (and incomplete) list of CAD domains, let us consider the example given in Figure 8-1. Here we have a four-link, two-slider mechanism. By defining the motion of the slider A in the form

$$x = Vt \qquad V = \text{const} \tag{1}$$

where

V = the speed of motion
t = time

we can calculate the motion of the slider B. Obviously, corresponding to Figure 8-1 we have

$$y(x) = \sqrt{l^2 - x^2} \tag{2}$$

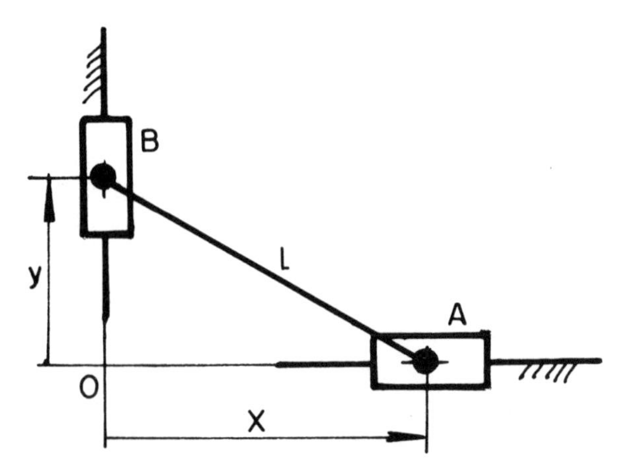

Figure 8-1. Four-link two-slider mechanism.

Expressing y in terms of time, we obtain from equations (1) and (2):

$$y(t) = \sqrt{l^2 - V^2t^2} \quad (3)$$

If this formula and the values of l and V are entered into the computer memory, the computer will be able to process the value of y for any time t. Of course, the speed and acceleration of the slider B can be obtained immediately by differentiating equation (3) once and twice respectively. It is important to note that this is not the only computation algorithm for this mechanism. For example, we can use the following expression for the definition of y:

$$\bar{y}(x) = \bar{x} \tan \varphi \quad (4)$$

where

$$\varphi = \arccos Vt/l \quad (5)$$

Thus

$$\bar{y}(t) = Vt \tan [\arccos Vt/l] \quad (6)$$

The programmer must decide which of the two algorithms is more convenient for the computer in question, will enable faster processing of the data and the like.

When the acceleration A_B of the slider B has been obtained, the force of inertia $F = A_B m_B$ can be found, and thus the load on the rod l can be determined. The rod's cross section can be calculated after computation of the stresses that the force F develops in the rod. We can choose to take into account only compressing stresses, or we can investigate, in addition, the ability of the rod to withstand buckling. For this purpose, another set of formulas, dependences, and constants must be introduced into the computer's memory.

An additional example of this type is the case described in Chapter 4 by the set of equations describing the model shown in Figure 4-12. A computerized analysis of this system would be of use in studying the behavior of the float on the waves and the

influence of the "springs" on the effectiveness of utilization of the wave energy. Computer-aided solutions are especially useful for nonlinear equations. In the case under consideration, for instance, when the shape of the "float" is conical (and not cylindrical), the force of buoyancy is essentially not proportional to the float's diving depth and thus equation (1), which describes its motion, becomes nonlinear.

Another kind of CAD problem is that of optimization. To illustrate this point, let us seek the minimal weight of a beam that can withstand a certain load (Figure 8-2). The weight can be minimized by changing the shape of the beam's cross section, within certain restrictions of the cross section's dimensions, and by changing the construction material. It is clear that here the moment of inertia I and the areas of the cross section are under discussion. The restrictions may be formulated as follows:

—The thickness of any element of the cross section cannot be less than δ.
—The height of the cross section cannot be more than H.
—The width of the cross section must be less than B.

By dividing the cross section into a number of elements, we can express its area of the following.

$$S = \sum_{i=1}^{k} s_i \tag{7}$$

where s_i = the area of the ith element
k = the number of elements

and the moment of inertia as

$$I = \sum_{i=1}^{k} I_i + \sum_{i=1}^{k} s_i D_i \tag{8}$$

where I_i = the moment of inertia of the ith element
D_i = the distance of the ith element's mass center from the symmetry axis of the cross section

The computer can be programmed to check different combinations of shapes and materials (the latter give us different values of allowed bending stresses and density of the material). The computer searches until it obtains the minimal values of beam weight that provide the required strength.

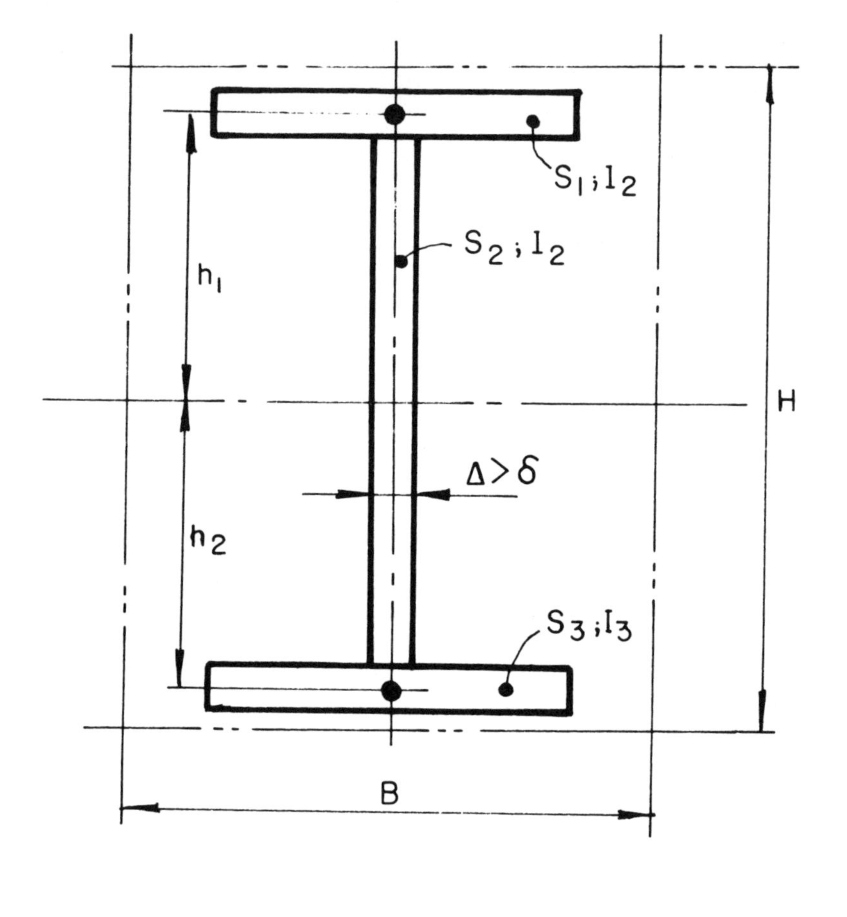

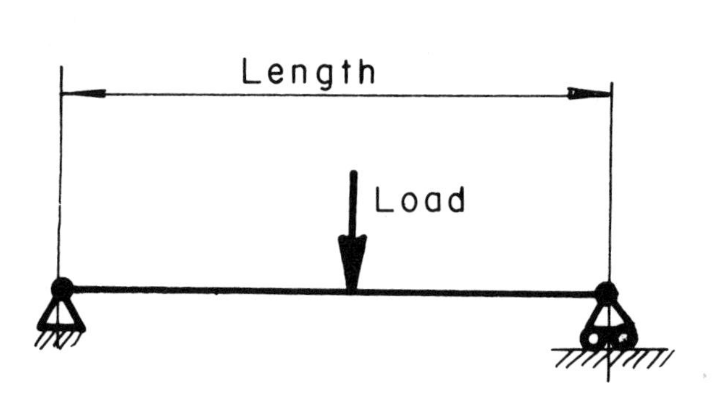

Figure 8-2. Example of a cross section of a beam.

If we introduce, in addition, a cost criterion (for the cheapest beam), the optimization will be more complicated. Figure 8-3 shows this phenomenon graphically. Here we see the "material" versus "cross section's shape" plane and above it the "cost" surface. The cost C is a function of the material and the cost of processing.

$$C = c(\text{material, processing}) \tag{9}$$

Thus the computer manipulates the numbers introduced into the memory in accordance with equations (7), (8), and (9). To accomplish such a design manually would require a tremendous

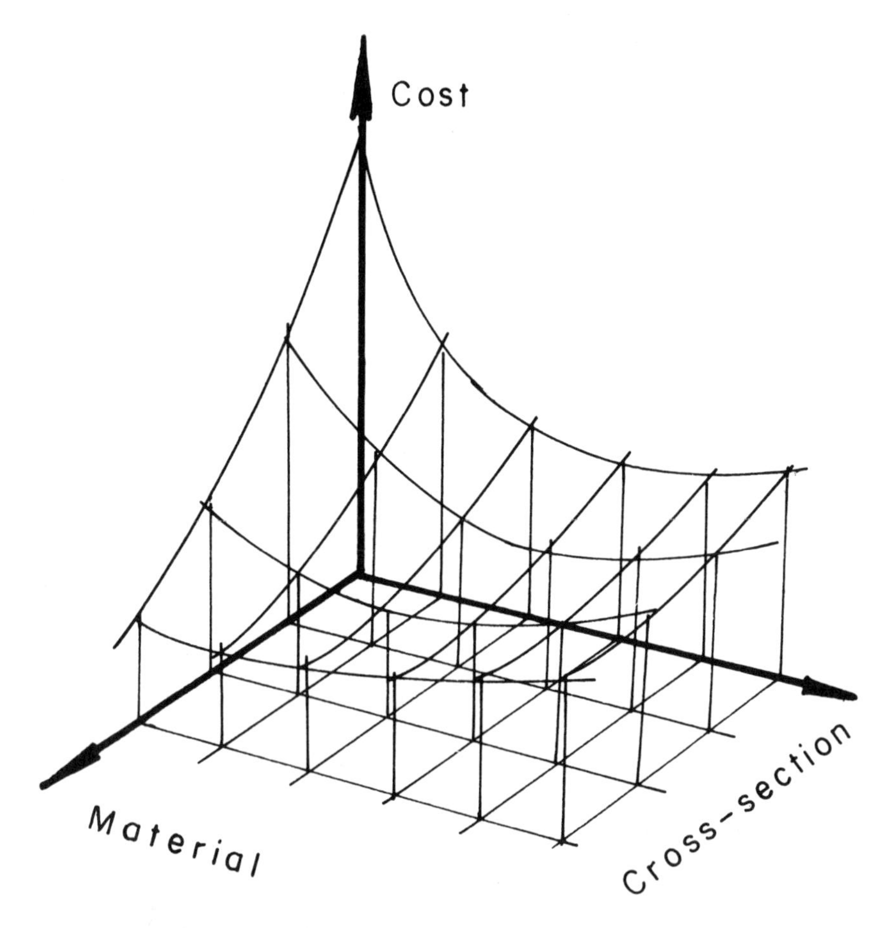

Figure 8-3. Graphic representation of a minimization surface.

amount of trial and error. The use of a computer facilitates the trial-and-error approach to produce the desired results incomparably faster.

By analogy, any other mechanical part—such as a shaft, gear wheel, key, or screw—can be designed in terms of known formulas developed analytically or experimentally. The speed of the computer facilitates the introduction of some very effective calculation methods, for instance, the finite elements method, which is based on relatively simple principles but requires a large number of repetitive computations. Only computerization of such methods can provide accurate and reliable results.

After the dimensions of the part or machine have been determined, the graphic stage begins. The computer parameters must be documented on paper for the manufacturer. The application of CAG at this design stage can be illustrated by the example of a gear wheel. The main diameters, the width, and the teeth modulus are determined by calculations and by the restraints imposed by the structure into which the wheel must fit, as shown in Figure 8-4. However, other dimensions can be chosen almost arbitrarily, such as those dictated by the manufacturing requirements and method or those that depend on the properties of the material of the part (in this example). The dimensions of the part may also be determined by the shapes and dimensions of auxiliary details, such as bearings or sealings. The convenience of CAG lies in the fact that the operator can manipulate proportions and dimensions on the screen until the optimal solution is found, and only then commit it to paper. Moreover, this graphic image and its program can be stored in memory and the design of other analogous wheels accomplished simply by pushing the relevant buttons.

Another application of CAG is in the graphic animation of images. The simplest way to illustrate this is again to use an example—the mechanism shown in Figure 8-1 modified to the view in Figure 8-5. (Point C traces an arc of a circle with a radius of one-half while point D traces a straight line dividing the coordinate angle.) If we want to know the trajectory of motion of some point belonging to a specific link, the animation technique is very useful. It is based on the analytic description of the coordinates of the chosen point. In other words, by substituting current values of time $t_1, t_2 \ldots, t_n$ in a suitable expression, we obtain, on the screen, the trace of the chosen point. By changing the values of sections

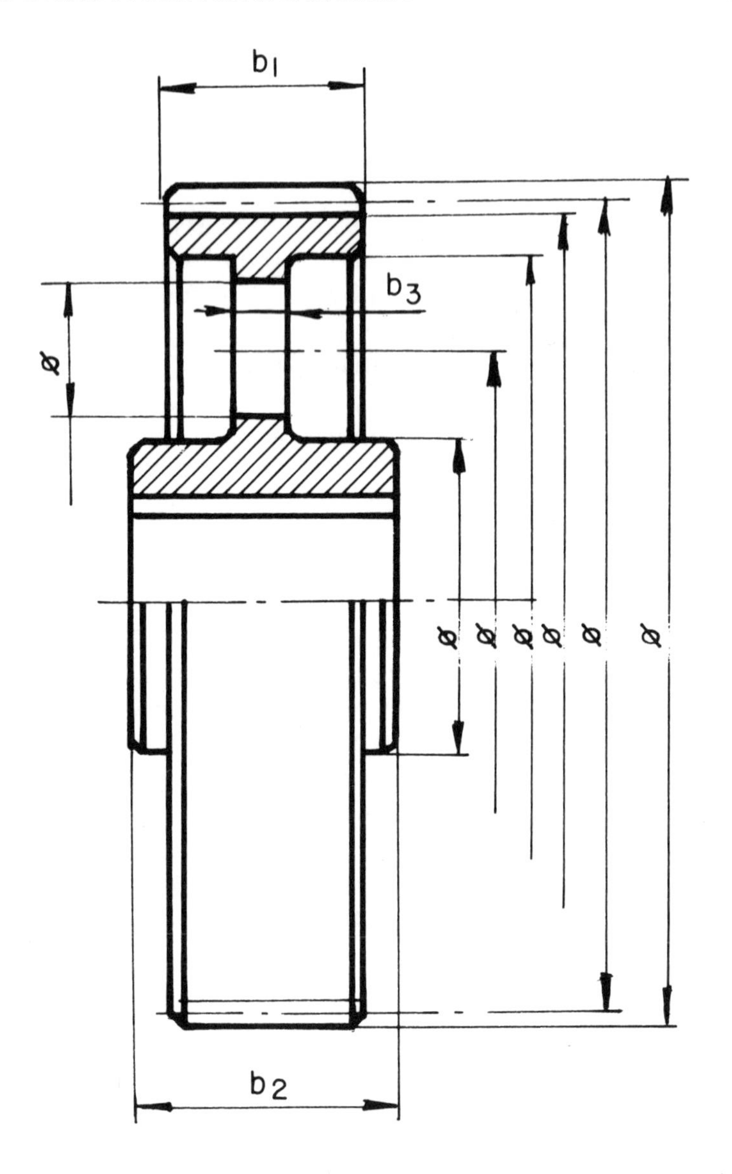

Figure 8-4. Gear wheel.

a and *b*, the operator can seek the proportions that provide the chosen trajectory.

Still another important application of CAG lies in the computer's ability to provide a spatial image of a particular structure by, as it were, rotating the structure in "space," thus allowing the operator to observe the design from different sides and different points of view. The time that can be saved by the use of CAG for this purpose in comparison with manual drafting simply cannot be estimated. Thus complex structures or structural elements with complicated shapes can be treated by CAG, and the influence of changes in the structure or elements can be studied with ease. An excellent example of the utilization of this CAG property is in the design of turbine or propeller blades, or of the bodies of cars or naval vessels. Obviously images are built on the screen by fast coordinate computing based on numbers and formulas or equations stored in the computer's memory.

CAM is the final stage in this chain of computerized activities. During this stage, a blank is transformed into a ready-for-assembly or ready-for-action detail. Only a limited number of manufacturing processes currently can be computerized, and there are some that are not suitable for computerization at all. To the former group belong material cutting and some assembly procedures. The latter group includes molding, stamping, and other kinds of processing based on plastic material deformation.

The principles constituting the basis of computerized manufacturing relate to the calculation of tool coordinates in accord-

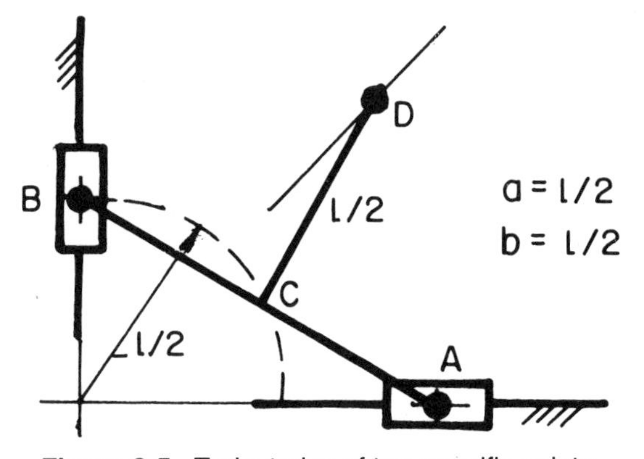

Figure 8-5. Trajectories of two specific points belonging to the connecting rod of a four-link two-slider mechanism.

ance with programs that the machine follows. In considering a computerized numerically controlled (CNC) milling machine, we can describe the motion of the cutter by some analytic expression or by a set or matrix of coordinates defining its location. The dimensions of the cutter can also be taken into account with ease. For example, to cut the profile given in Figure 8-6, the coordinates of the arc centers O_1, O_2, and O_3 must be defined, as must the radii R_1, R_2, and R_3 of the arcs. Then the current coordinates of the cutter's center O_4 are calculated while the coordinates x and y must satisfy the obvious expression:

$$
\begin{aligned}
(x - x_1)^2 + (y - y_1)^2 &= (R_3 - r)^2 \\
(x - x_1)^2 + (y - y_1)^2 &= (R_2 + r)^2 \\
(x - x_2)^2 + (y - y_2)^2 &= (R_1 + r)^2 \\
(x - x_3)^2 + (y - y_3)^2 &= (R_3 - r)^2
\end{aligned}
\tag{10}
$$

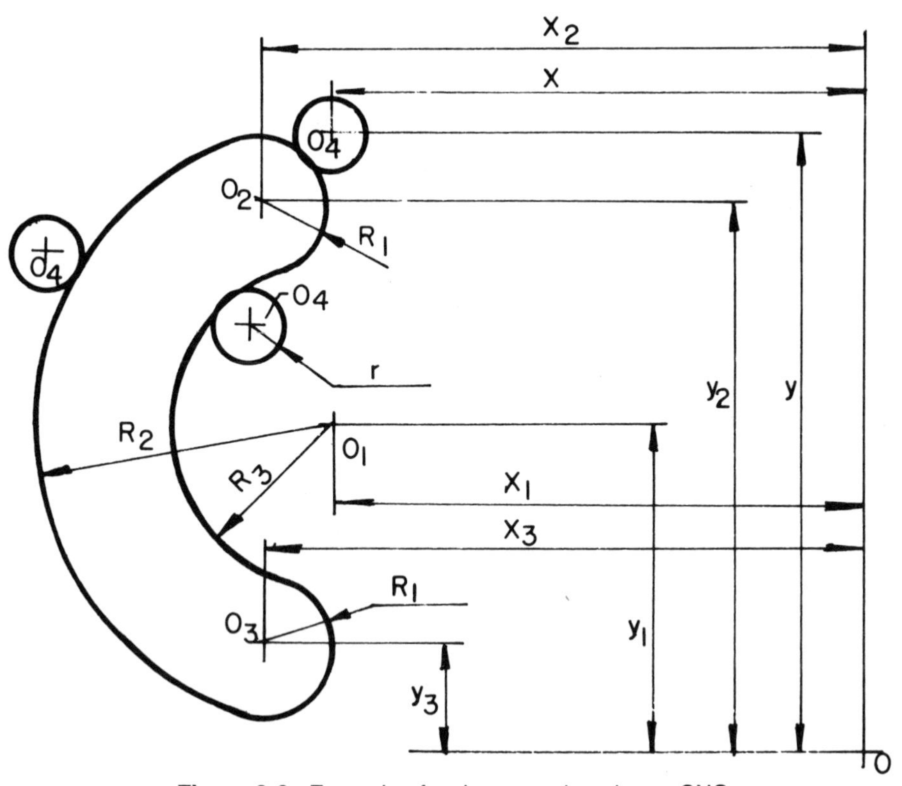

Figure 8-6. Example of a shape produced on a CNC milling machine.

In addition, such a CNC machine can change tools in accordance with previously defined requirements and cutting regimes to provide optimal surface quality in minimal time and with minimal tool wear.

An electronic circuit assembly machine can be analogously programmed. The components (capacitors, resistors, transistors, etc.) that must be positioned are described by codes, and the plate on which the assembly is carried out is positioned in accordance with a coordinate matrix loaded into computer memory. The system can be improved by programming the computer to calculate the shortest trajectories and minimal times of the positioning mechanism. Such improvements obviously increase the efficiency of the process.

Thus we see again that despite the fact that the task that CAM must achieve is physically different from the aims of CAD and CAG, the underlying principles are similar.

The main conclusion that we can derive so far is that any computer-aided activity must be based on:

1. Known mathematical models (formulas, equations, and inequalities) for CAD, CAG, and CAM.
2. Known sets of constants and parameters (coefficients of friction, stiffness coefficients, specific weight, safety factors, allowed stresses, etc.) for engineering purposes.

CAN THE COMPUTER BE CREATIVE?

"The human is smart, creative, and slow. The computer is stupid, noncreative, and fast."[12] In other words, the computer does not possess such human attributes as curiosity, imagination, intuition, and the ability to take the initiative; it does not know how to formulate questions. For this reason the child who is asking, "How much is 3 plus 2?" is much more creative than the computer, despite the fact that the computer "knows" how to solve differential equations. The behavior of a computer is instinctive rather than creative: it follows blindly the program introduced into its memory. From this point of view, its thought pattern is similar to that of an insect or a bird. For millions of years, the seasonal

migration of birds has been controlled by an unknown navigation technique. A variety of explanations have been offered for this phenomenon. Some experts feel that visual markers such as seashores are the guides that direct the flight path; others believe the sun or the magnetic field of the earth serves this purpose. The point we wish to make here is that birds are able to find their way without an understanding or a knowledge of astronomy and geography, and without the ability to compute a course. (The sad truth is that those who lose their way die.) When, however, faced with a nonstandard situation, birds and insects become completely "illiterate." Modern science explains such phenomena in terms of innate instincts, genetic codes, or inherent abilities.

The rigid behavior of a migrating bird is reminiscent of that of a computer. For example, a conventional CNC milling machine will continue its cutting process (as shown in Figure 8-6) blindly even if it encounters a hard grain (say, in the cast iron) accidentally present in the material, with the outcome being dictated solely by good or bad luck. Either nothing detrimental will happen (if the grain is small and weak), or the result may be a broken tooth on the cutter, a spoiled part, a slip in one of step motors, or the loss of control. Or let us take the example of a programming error. If in the calculation of the moment of inertia of the beam (Figure 8-2) the expression had accidentally been written incorrectly (say, instead of $I = bh^2/12$, the wrong $I = bh^3/12$), the computer would have continued processing the data without hesitation. How would a creative creature respond in such situations? We can suppose that upon feeling a growing resistance to the cutter's blades, such a creature would stop cutting and start analyzing the source of the resistance, and then derive conclusions and act accordingly—perhaps by decreasing the cutter's speed or changing the blank. In the second example, the error would be revealed practically immediately after the first few results had been obtained, since an experienced engineer would intuitively feel that the values obtained were greatly disparate from the anticipated results. Of course, a torque meter or sensor could be installed in the milling machine and some computerized checking routine used to stop the machine's activity, but such measures are effective only when possible problems can be predicted. The computer is not able to deal with failures that have not been predicted by the programmer.

In contrast to an insect, or a computer, human beings have inherited the ability to handle *nonprogrammed, nonstandard situations*, as well as to create ideas, tools, and concepts. Take the case of the famous actor who had forgotten to darken his arm when appearing in the role of Othello. The astonished audience reacted with whispers, smiles, and laughter as the actor continued his performance. During his next appearance, the audience was even more surprised when, at one point, the actor pulled off a white glove and exposed a dark-colored arm. Only the human brain is capable of this type of creative behavior.

Let us try to understand what makes a certain action creative, whereas another, though much more complicated, is not creative at all. Why should a child at play putting one block on top of another be creative, but a computerized manipulator, a so-called robot, assembling a complicated electronic circuit be considered noninnovative? We do not purport to analyze the nature of creativity or how it can be related to artificial intelligence, but we do intend to examine computer-aided technical creativity. We thus must define where the boundaries of creativity lie.

To return to the playing child, we can ask: "What makes the child creative or innovative?" It is the ability to see in the pile of blocks a tower or a building and to associate it with other things and events. In other words, the capacity for imagination and for formulating and asking questions—the child's curiosity—are the main features determining creativity.

When James Watt encountered a fuel-consumption problem with the steam engine, he responded by questioning what could be done to rectify the waste of fuel. He answered this question by improving piston sealing, applying steam from both sides of the piston, and exploiting the advantages of compound engines. He also created a rotating output and a centrifugal speed governor. Finally, he invented a device that graphically indicated the pressure inside the engine cylinder and then used the device to monitor the performance of the engine and to adjust it to facilitate maximum efficiency.

Some people seem to be born with the ability to play chess well, to "feel" a good move. The chess-playing computer must overcome by blind exhaustive search what the human brain is able to "jump over" by intuition, by inspiration, or by using rules of thumb. Concepts such as esthetics or a sense of elegance are

foreign not only to a computerized chess player, but also to a computer-aided creator in any field.

Let us return to the delineation of creativity. We can say that creativity begins when a system (electronic or live) is able to initiate a question or recognize a problem and then devote itself to seeking an answer. The pessimistic view of computerized creativity is that the computer can never be anything other than a high-speed "number cruncher" (it knows how to handle zeros and ones, and it can carry out simple operations such as addition and subtraction millions of times per second). In addition, the computer does not have the ability to make chance discoveries. For example, the computer could not have discovered radioactivity, as did Henry Becquerel when he observed in 1896 that certain salts of uranium were particularly active in spoiling photographic film. The computer also could not have been capable of discovering Newton's law of gravity. Richard Bellman[11] has said: "At present, and most probably forever, we cannot use a computer to recognize structure. However, there is no proof of this conjecture and it may well be possible that tomorrow someone will find a way to use a computer for this purpose."

If computers have such limited capabilities, one might ask, then why continue the search for applications? Computers do have positive features, and a look at these will indicate what awaits us in the field of computer-aided creativity—at least how the computer can be applied in the preparatory stages of design, the preliminary draft work.

The speed at which the computer works is but one of its advantages. The human brain often relies on feelings, while the computer, in a blind, stubborn, but very fast way, may attain the same result. In some instances a "good human brain" may shorten a search by intuitively rejecting deadlocks, but often will wander aimlessly among blind alleys and take many years to find the way out of scientific or technical impasses. This psychological inertia is one of the most serious obstacles to human creativity, especially in applied domains and technical fields. Perhaps computer-aided creative activity will suffer less from psychological inertia (or stereotyping as it is also known). For good or for bad, the computer is not influenced by tradition, and so will not be "embarrassed" to find and to offer an unusual or unfashionable solution. A computer-aided creativity (CAC) system would not have been sur-

prised or felt unsure when offering a front-wheel-drive concept for a car, despite the fact that rear-wheel drive was the convention, or when proposing a circular-arc tooth profile instead of the usual involute or cycloidal profile. A computer could have proposed micromini bathing suits years ago because they would have appeared functional, convenient, and healthful and the computer would not have taken into account norms or prejudices.

Another advantage of CAC systems over human thinking results from the fact that the computer "deals" with collective knowledge. Here we come to the requirement for limiting the volume of the computer's memory. In principle there are no limits to the size of the memory and the amount of information that can be stored. The bigger the volume of stored knowledge, the closer it approaches the so-called ideal. (Ideal knowledge is based on the supposition of an infinitely large memory.)

To be creative a system must possess some knowledge or information. There are two ways of storing information. The first, the direct, way consists of tables or matrixes of numbers. For instance, specific weights of different materials can be simply introduced in the form of a two-dimensional table of friction coefficients, shown in Table 8-1.

The other way to store information is based on the generation of information, that is, the memory is loaded only with the analytic expressions that are used to compute a required number in accordance with required conditions. For instance, the cross-sectional areas, static moments, and moments of inertia or round rods produced by a particular company can be calculated at any time, in accordance with stored formulas, by feeding the computer only the diameters of the rods.

Another advantage of the computer is that it does not become tired (even without coffee breaks). It is able to continue searching where the human brain would be saturated and require rest. In addition, the computer's memory does not forget the input data or, more important, the results obtained.

Thus there is a spark of hope that the advantages of the computer can be applied to creativity, and efforts to attempt to get computers to simulate intellectual activities may be worthwhile. This explains the appearance of the relatively new concepts of artificial intelligence, computerized heuristics. In the field of medicine, for example, computer-aided solution search has already

been successfully applied. The question is: How can computers be utilized best in the domain of engineering and technical creativity?

COMPUTER-AIDED CONCEPTUAL DESIGN

We already have discussed some aspects of computer-aided creativity and have seen that this is a very complex problem. Nevertheless, in spite of the limitations imposed by machine-design conceptual constraints, we hope to find some justification for

Table 8-1.
Example of Organization of Information

	Material 1	___ 2	. . .	___ i	. . .	___ n
Material 1	f_{11}	f_{21}	. . .	f_{i1}	. . .	f_{n1}
___ 2	f_{12}	f_{22}	. . .	f_{i2}	. . .	f_{n2}
. . . . . .	. . .	. . .	. . .	. . .	. . .	. . .
___ i	f_{1i}	f_{2i}	. . .	f_{ii}	. . .	f_{ni}
. . . . . .	. . .	. . .	. . .	. . .	. . .	. . .
___ n	f_{1n}	f_{2n}	. . .	f_{in}	. . .	f_{nn}

Note: f_{in} is the friction coefficient for the pair of materials number i and number n.

the development of algorithms useful in creative machine design. The main problem in introducing the computer into the design stage is the lack of mathematical models describing the solutions. Previously, we saw that any computer-aided activity in engineering is based on the existence of a more or less accurate mathematical description of the phenomenon under investigation. Galileo once said that "mathematics is the language of science"; this is also true for computerization. We must find some way to formalize the processes used in conceptual thinking. But because we do not know what human thinking really is, we cannot try to copy this process. (And, as we showed, earlier, directly copying nature is not always the best way to proceed.)

In this section we show that some computer-aided means of releasing human energies for more interesting and difficult activities are indeed available (although we doubt that there is anything more fascinating than finding new solutions). Algorithms are not necessarily intended for finding high-level solutions; some algorithms can provide computation results that enable the user to make the final decision. Thus the system under consideration really illustrates the interaction of the person with the machine.

The computer deals with binary numbers, with zeros and ones, and so can be used effectively for handling logical operations. As an example, let us formulate two attributes of some set of technical entities:

$$T_1 = \text{high flying speed}$$
$$T_2 = \text{high take-off speed}$$

The 0 denotes the absence of a property, and the 1 the existence of that property. Thus we have Table 8-2.

Table 8-2 automatically gives us a list of all the technical possibilities that relate to the chosen attributes:

0. Slow take-off and slow flight.
1. Fast take-off and slow flight.
2. Slow take-off and fast flight.
3. Fast take-off and fast flight.

Analysis of this list brings us to the idea (case 2) of creating an aircraft (an answer to the "what" problem, but not "how" to

achieve it) that does not require long and expensive runways but still achieves high cruising speed. Thus if we load the computer with k attributes, it offers us 2^k combinations. We then provide the computer with information for which combinations already exist, and perhaps also with information as to which combinations do not interest us, and we obtain a list of the remaining possibilities. In our example, case 0 (a helicopter) and case 3 (a conventional airplane) are known, and case 1 seems ridiculous. Therefore, case 2—a brilliant solution of a plane with rotatable wings or engines—is ours! (Here we have answered the "how" question.)

Another very powerful technique that can be applied in computer-aided creativity is the so-called experts' judgment. The computer's memory can store experts' estimations of methods, approaches, shapes, techniques, properties, and so on. As mentioned, one of the advantages of the computer lies in its ability to collect information, knowledge, and experience from many sources. In addition, it is capable of processing this information and producing an average estimation. For instance, we may ask how to produce an opening L millimeters in length and D millimeters in diameter in a specific material. A program tailored to answer such a question would use the recommendations of a number of experts (which do not always coincide). The program would carry out a sort of "voting" procedure, finally recommending what the majority of experts would have recommended. Let us take another example. Say we have to decide what material is suitable for a static structure such as the housing of a ball bearing (where weight is not a particularly significant consideration) of which about 1000

Table 8-2.
Matrix Describing the Set of Entities

	T_1	T_2		
S_1	0	0	0	
S_2	0	1	1	
S_3	1	0	2	decimal numbers
S_4	1	1	3	

units a year will be produced. Most engineers would choose cast iron. This represents the kind of problem, but more complicated in nature, that can be solved by a computer.

To illustrate the next avenue of thinking formalization, we use a historical example. Thomas Young (1773–1829), an outstanding scientist in his day, is famous for his exposition in 1807 of a mechanical constant known as Young's modulus. This constant is used in the well-known equation describing the elasticity of material:

$$\sigma = \varepsilon E \qquad (11)$$

where

σ = the stress developed in the material under tensile force
ε = dimensionless elongation, which is defined as the ratio between the elongated pattern and its initial length
E = Young's modulus

We do not know exactly how Young carried out his experiments, but he would have had to invoke the following type of procedure. He would have prepared identical shapes from different materials, and then stretched the shapes by applying increasing forces and simultaneously recording the change in the lengths of the shapes. In dividing the new length L by the original L_0, he would have calculated ε:

$$\varepsilon = L/L_0 \qquad (12)$$

Then he would have built the dependence between the applied force F and ε. We assume that the area of the cross section A of the shapes did not change during the elongation, and therefore the stress σ is proportional to the applied force F:

$$\sigma = F/A \qquad (13)$$

Young then would have possessed the group of dependences that enabled him to derive the following conclusion: In the range of elasticity, every material under stretching load changes its length

in proportion to the applied force and is described by a specific constant or proportionality coefficient E.

Now let us imagine that groups of numbers gathered from these experiments were introduced into a computer programmed to look for analytic approximations. The computer would reveal very quickly that the best approximation is a linear one and that the proportionality coefficients for each group equal a certain constant. If we give some thought to this example, we see that the human brain can be stimulated by information collected from observation or experiments (as in the case of Thomas Young). On the other hand, the computer is not able to initiate (at least for the time being) a search since it does not possess the curiosity to analyze the numbers; however, as we see from the example of Young's modulus, it can simulate the process of analysis. In such a case, the computer operates quickly and accurately, and can estimate the reliability of the "discovered" laws. We must emphasize here that the computer has been programmed for this type of discovery. For technical purposes, we would consider it a satisfactory achievement if the computer could solve creative problems on the level of Young's law.

To enable the computer to approach creative problems, therefore, we have to feed the memory with information and certain rules to handle it. This requires a coding or language. We mentioned earlier that the creative, conceptual stage of design cannot be described in terms of mathematics, but because of the nature of the computer, cooperation with it can be built only on a mathematical basis. We thus have to find formulas, equations, and algorithms imitating the processes of technical creativity. In other words, we have to furnish the computer with rules and strategies in a form it can understand. This formalization of creativity in the field of engineering is the task of those who want to mobilize the computer's power in the search for technical concepts and innovation. Of course, we cannot expect a comprehensive computerized answer to problem solving to appear in the near future. The creation of such a program is not a routine operation, and is more of an art than a science.

Let us consider one more illustration of a possible formalization method—this time in the "how" category. Let us say the problem is to find an optimal solution for a device with definite kinematic

properties, for instance, a planar mechanism providing a linear reciprocating output for a rotating input motion. For creative problems of this type, the computer's memory must first be supplied with the available relevant knowledge. (This step must be taken before the question arises, and constitutes a "library" that is continuously enriched. Thus when using it for a particular problem, one is not aware of the content of the memory.) The type of information that would be stored in the computer's memory is shown in Table 8-3:

Table 8-3

Section of Stored Information

i	B_i	Figure no.
1.	Four-bar linkage	8-7
2.	Cam mechanism	8-8
3.	Gear transmission	8-9
4.	Belt drive	8-10
5.	Frictional drive	8-11
6.	Crankshaft-slider mechanism	8-12
7.	Noncircular wheel gear transmission	8-13
8.	Rack and wheel	8-14
9.	Cylindrical pneumomechanism and	
10.	Cylindrical hydromechanism	8-15
i		
n		

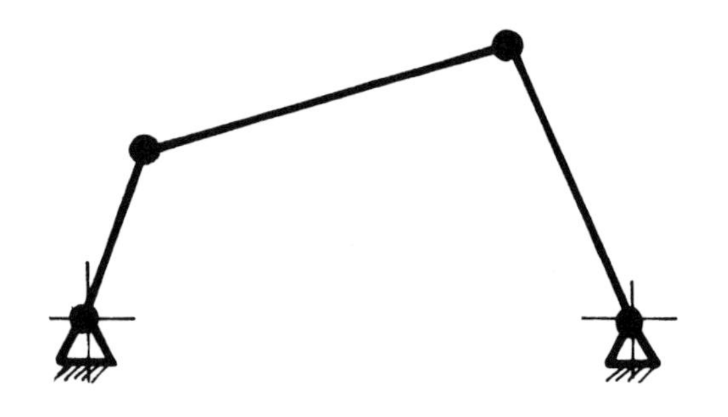

Figure 8-7. Four-bar linkage (See Tables 8-3 and 8-4.)

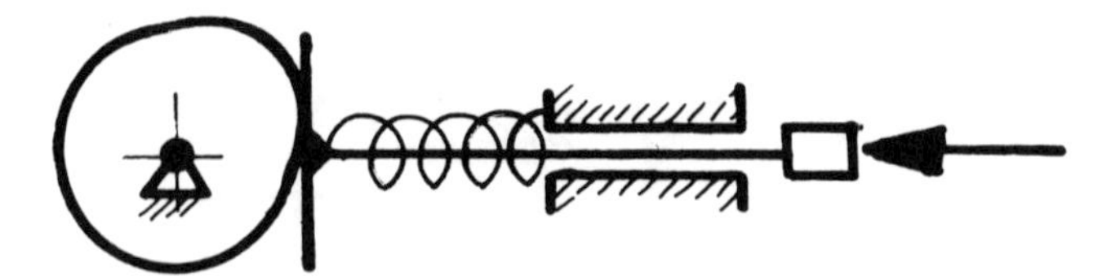

Figure 8-8. Cam mechanism (See Tables 8-3 and 8-4.)

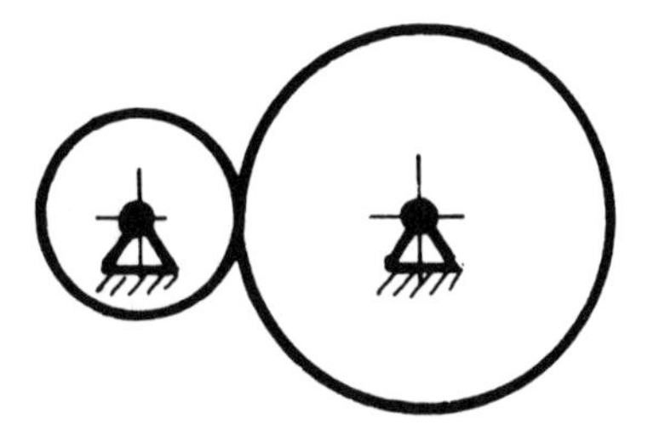

Figure 8-9. Gear transmission (See Tables 8-3 and 8-4.)

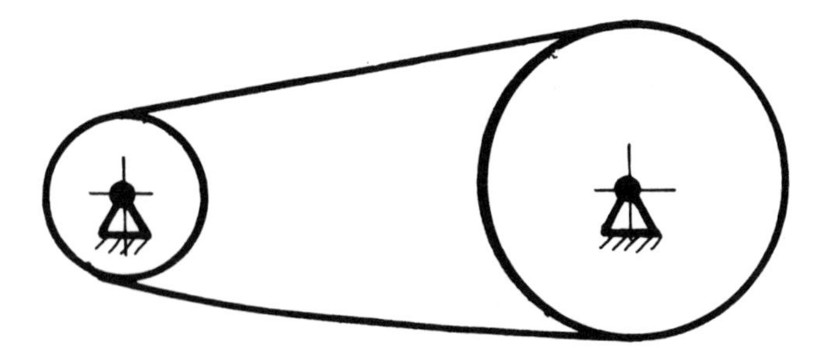

Figure 8-10. Belt drive (See Tables 8-3 and 8-4.)

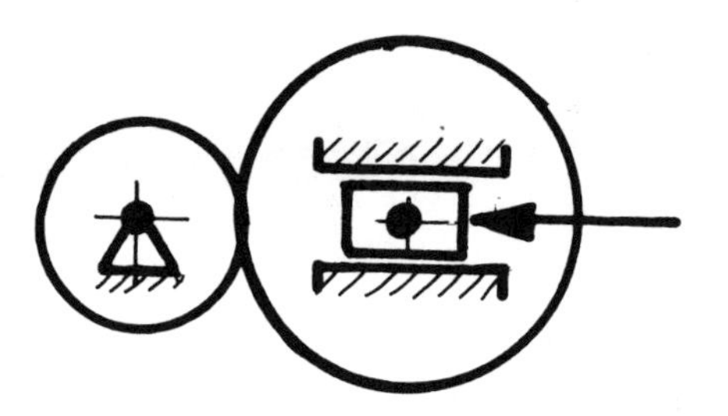

Figure 8-11. Frictional drive (See Tables 8-3 and 8-4.)

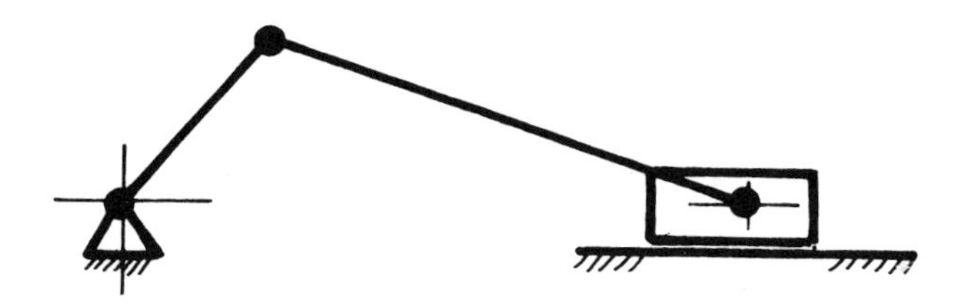

Figure 8-12. Crankshaft-slider mechanism. (See Tables 8-3 and 8-4.)

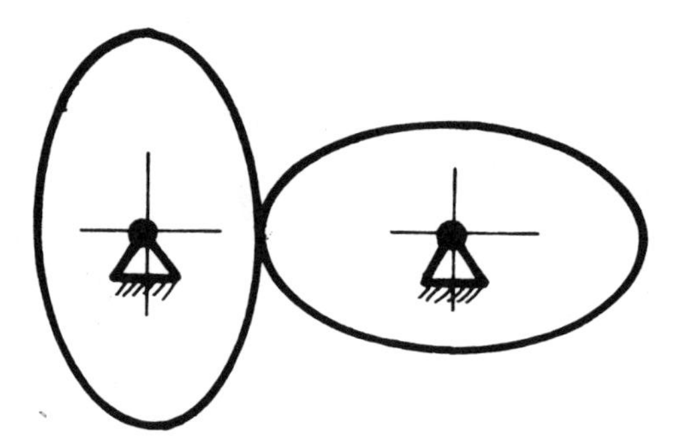

Figure 8-13. Noncircular wheel gear transmission. (See Tables 8-3 and 8-4.)

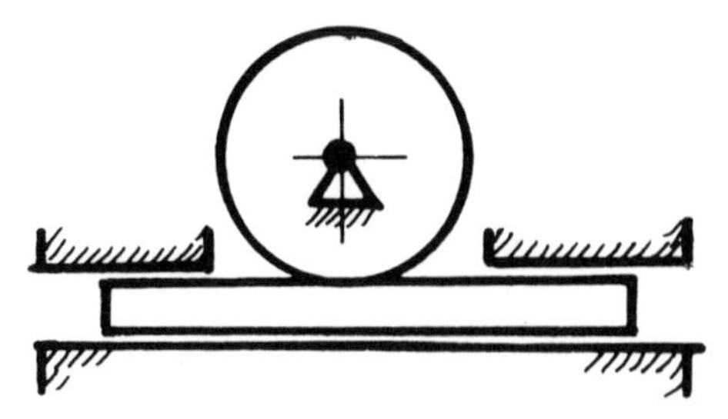

Figure 8-14. Rack and wheel (See Tables 8-3 and 8-4.)

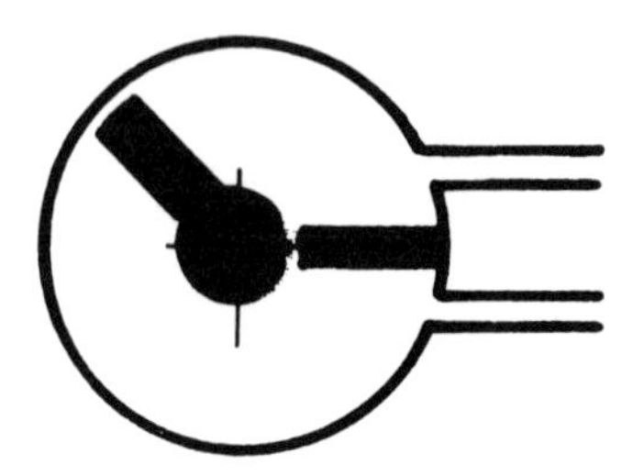

Figure 8-15. Cylindrical pneumomechanism and cylindrical hydromechanism (See Tables 8-3 and 8-4.)

Table 8-4.
Set of Atributes Describing Subjects in Table 8-3

j *Q*

1. *The driving link rotates*
2. *The driving link moves with an angular motion*
3. *The driving line moves with a linear motion*
4. *The driven link rotates*
5. *The driven link moves with an angular motion*
6. *The driven link moves with a linear motion*
7. *The mechanism is planar*
8. *The mechanism is spatial*
9. *Discontinuous motion*
10. *Continuous motion*
11. *Interrupted motion*
12. *Constant ratio*
13. *Periodical ratio*
14. *Nonlinear ratio*

j

k

In Table 8-3 each line represents an entity B_i *that potentially can be a sought-after solution. The proposed algorithm requires a quantitative description of each entity. For this purpose a set of properties* Q_j is introduced (Table 8-4).

We now have *n* entities described by *k* properties or attributes. Each line in Table 8-3 thus can be stored in the computer's memory by a *k*-dimensional binary vector because the simplest way of formalizing this information is to write it in the following manner:

$$\bar{B}_i(q_{i1} \cdots, q_{ij} \cdots, q_{ik}) \qquad (14)$$

where

$$q_{ij} = \begin{cases} 0, \text{ when the property } Q_j \text{ does not exist} \\ 1, \text{ when the property } Q_j \text{ does exist} \end{cases} \qquad (15)$$

Upon organizing this information, we obtain Table 8-5.

Therefore, in the computer memory the information describing all known mechanisms appears as a matrix, a fragment of which is shown in Table 8-6.

Table 8-5.
Organization of Information from Equations (14) and (15)

	Q_1	. . .	Q_j	. . .	Q_k
B_1	q_{11}	. . .	q_{1j}	. . .	q_{1k}
. . .	. . .	. . .	. . .	. . .	. . .
B_i	q_{i1}	. . .	q_{ij}	. . .	q_{ik}
. . .	. . .	. . .	. . .	. . .	. . .
B_n	q_{n1}	. . .	q_{nj}	. . .	q_{nk}

Obviously the sought-for solution C can also be described as a binary vector in the following form.

$$\bar{C}\ (P_1, \cdots, P_j, \cdots, P_k) \tag{16}$$

And here also

$$P_j = \begin{cases} 0, \text{ when the property } Q_j \text{ is not required} \\ 1, \text{ when the property } Q_j \text{ is required} \end{cases} \tag{17}$$

The algorithm must find those known and stored solutions that best fit the required conditions. What the algorithm does can be expressed as follows:

$$\bar{S} = \bar{B}_i(q_{i1}, \cdots, q_{ij}, \cdots, q_{ik}) - \bar{C}(P_1, \cdots, P_j, \cdots, P_k) = 0 \tag{18}$$

To describe equation (18) in words, we subtracted the sought-after value from each of the stored values and ordered the computer to print out only those i numbers (remember, i is the number of a line in Table 8-5 that corresponds to a specific mechanism)

Table 8-6.
Matrix Fragment

```
10001010010010
10000110010010
10010010010100
10010011010100
10010011010100
10000110010010
10010010010010
01000111100010
00001110001000
00001110001000
```

that promise complete identity of properties between the stored and sought-for mechanism. The mechanism we are seeking can be formally described in terms of equation (16) in the following form.

$$\bar{C}\ (P_j) = 10000110010000$$

Thus the operator obtains several different solutions for the general case. In our specific example, these solutions may be

1. Crankshaft-slider mechanism.
2. A cam and slider-follower.

Lines 2 and 6 fit this description, and from Table 8-3 we find the "unknown" mechanism.

EXERCISES

1. Try to "invent" new machine tools imitating the creativity process with the aid of the algorithm shown in Table 8-2 for the following list of attributes.

T_1—the blank rotates
T_2—the blank moves linearly
T_3—the cutter rotates
T_4—the cutter moves linearly

2. Continue Table 8-3 for about another ten entries and describe them in binary form using the attributes given in Table 8-4.

3. What mechanism will answer the sought vector $\bar{c}$ described in the following form?

10010010010100

Describe it in words in accordance with Table 8-4.

Chapter **9**

MARKETING

We have discussed the technical aspects of the design process of new engineering-based products, machines, and devices. There is, however, a wide field of activity that is devoted to introducing the engineer's creation to the consumer. This activity is aimed at producing income for the manufacturer, not only as profit, but also to compensate the manufacturer for expenses incurred in the development of the new product and to facilitate its further improvement. At this stage salespeople rather than engineers are usually mobilized to deal with market penetration. The marketing problems involved are based on:

* The commercial situation.
* The psychological and intellectual orientation of the consumer society.
* The political climate.
* The technological level of the society.

Thus the engineers turn over the articles which they have created to salespeople for introduction to the marketplace. Obviously there must be some interaction between these two groups of workers.

THE PROBLEMS

In this section some ideas pertaining to the marketing aspects of design are considered briefly. It should be noted that, to some extent, the problem of marketing should be dealt with in the

"what" and "how" solution-seeking directions. When a new product, machine, or device (a new "what") is introduced into the market, the problems differ from those that arise when a new model of an existing product (a new "how") penetrates the market.

For a product, machine, or device to be competitive, it must be characterized by one or more of the following.

* It must be cheaper—which implies lower costs of production, exploitation, and installation.
* It must be more functional—that is, be more efficient, more accurate, more reliable, safer, more productive, and so on.
* It must be more esthetic—that is, be more pleasing in form, color, texture, and so on.
* It must be more easily exploitable—that is, be more convenient to use, smaller, lighter (if desirable), and so on.

If the company is already in the market to which the new product belongs, it is, of course, in a better competitive position. Otherwise it must offer something that will surpass the merits of competitors' products to a superlative degree. From this point of view, "how" solutions are pertinent to those manufacturers already in the relevant business. Newcomers to a particular marketing branch must come out with new "whats" if they intend to penetrate the market.

It is of great importance to distinguish between inner and outer markets; that is, the volume of the market plays a significant role in decision making. This is especially true for small countries. No less important is the geographical location of the market. What is worthless in one place can be sold in another. For instance, engineers working on a chlorination device for swimming pools found an original solution for chlorination of the water. The question then arose as to whether it would be possible to apply these ideas to water chlorination in general. After a market search had been conducted, it was concluded that in some underdeveloped countries that lack central water-purification facilities, the new device could be installed in private homes to chlorinate well water for domestic consumption.

This is an example of a product finding a market because of the relatively low technical level of the country. It would be a

mistake, however, to assume that any product could be sold here. Care must be taken not to attempt to introduce "overqualified" products into this type of market: the lack of service for the product, machine, or device, and the relatively low technical expertise of the users will combine to make the introduction a failure. For example, one cannot sell modern automobiles in a country with no modern roads (unfortunately such societies often purchase tanks and the like).

Thus the questions the designer must answer before a technical solution can be offered for marketing are not solely of a technical nature. A wide range of commercial and juridical considerations also must be taken into account before the decision can be made as to whether the product will succeed in the market. To this end a substantial market research effort is usually devoted to determining the place of the new product in the market. Of course, it would be preferable to have a complete picture of the market before efforts to create a new product are even initiated, but in practice it is very often the case that only after a product or solution has been created is enough information available for conducting market research.

When a market search is carried out, the time factor must always be kept in mind. To develop a machine, device, or product, time is needed. Time is also needed to investigate the market requirements, the commercial aspects, and the technical specifications. As a result of these seemingly banal facts, the engineer often is successful in creating a product, only to find that it has little or no commercial effect.

THE CONCLUSIONS

From the foregoing discussion, we can draw the following conclusions:

1. To save time market research and product development should be carried out to some extent in parallel.
2. It is sometimes better to patent a nonperfected product than not to develop it at all.

These conclusions can be illustrated by the following example. Engineers in the author's company came up with the idea of developing a device that would warn drivers when they exceeded the speed limit, and thus help ensure a safer ride. Because we realized that such an idea was too simple to be original, in parallel with the beginning of the development of the device, we began to investigate the market. We knew that the simplest technical solution would be mechanically to link the speed indicator of the warning device to the rotating parts of the car engine or transmission. Thus to increase the potential for marketing the product, we decided to avoid using mechanical connections. We felt that this approach would offer the following advantages.

1. The possibility of inventing something already invented would be less.
2. The buyer would have no problem in installing the new product (a convenience property of the product).
3. From the previous point, it follows that the general cost to the consumer would be lower.
4. The device would be more easily adaptable to different car models.

Most of the competitive devices already available on the market, and those that appeared at about the same time (about 1980), were designed to be inserted between the speedometer of the car and the flexible cable shaft. Another type of competitive device included an adapter that had to be fastened to the ignition distributor shaft. The latter had the disadvantage that the gear ratio influences the operation of the device.

Our approach led to the birth of a device—called "the speed warning bleeper"—which, although not the only one on the market, was competitive with the others, because the time factor had been taken into account during its development.

Another example was the invention of the carbon lamp in 1878–79 by Thomas Edison (1847–1931). He also designed and manufactured all the accessories required for the introduction of his lighting techniques. His invention was, however, a failure because it was based on the use of direct current. During the same period, Edison's competitor Nikola Tesla (1856–1943) invented the much more efficient polyphase alternating-current system. A

power struggle ensued between Edison and Westinghouse (who bought the rights to Tesla's patent). Of course, this quarrel impeded the technical progress of both parties to a certain extent. Modern engineering history abounds with such collisions. Engineers often have to design an entirely new product rather than improve an existing article to circumvent competitors' patents. In addition, an original patent is sometimes bought, but never exercised, for commercial reasons. For example, even though the new product might not have great intrinsic value, it might affect the profit of a competitive item. In this case purchase of the rights to the new patent stops the "intruder" from penetrating the market. This is another example of marketing policy that has nothing to do with technical solution seeking. This aspect of marketing will not be discussed here. We will deal only with the creative and technical aspects of marketing problems (while keeping the political aspects in mind).

We view the first step in this direction as the market study. There are a number of examples that show that patent laws and commercial interests may prevent optimal technical solutions from penetrating the market in the best way. We will consider two of these.

The first is a well-known story. James Watt (1736–1819), the inventor of the steam engine, could not use the simple, reliable slider-crank mechanism for transmission of the reciprocating piston motion into rotation of the crank and the flywheel. This widely known (and today much used) mechanism had already been patented by somebody else. Watt created another four-bar linkage to guide the piston rod of his engine on an approximately straight path, and in 1781 he invented the sun-and-planet gear for generating the rotating motion.

The second example is taken from our own experience. Before finalizing the design of a flowmeter and commencing its manufacture, we performed a market analysis. Some of the conclusions we reached as a result of this analysis are used in the following to illustrate what a market search is all about.

It appeared that:

1. There are three main fields of application for flowmeters for low flow rates (such as the type we developed); these are, in order of preference:

a. Medical uses.
b. Measurement of fuel consumption.
c. General uses in chemical and other laboratories.

2. The flowmeters already on the market designed for a flow range of about 0.2–10 cm^3 sell for about $200–$250, whereas the price of our device was estimated at about $30.

3. The design concept we used for our flowmeter had certain advantages over existing competitive designs.

4. Medical applications require far higher sensitivity than we could offer (thus, despite the fact that this domain could constitute the biggest market, we had to withdraw from the medical field).

5. Flowmeters of the type under consideration must be able to sustain a certain internal pressure (about 20 atm).

This market analysis, together with a working prototype, was presented to potential manufacturers. Our offer was accepted by a modestly sized company involved in the fluid and gas-flow measuring business. The subsequent turn of events constitutes a good example of the difficulties that accompany the process of introducing a new product into the market.

The engineers of the accepting company redesigned the flowmeter so that it could be manufactured by using metal- and plastics-casting techniques (the prototype was designed to be produced by cutting). On the basis of the amount of material that was needed and the production costs involved, it was estimated that the selling price could be fixed at as little as $25 per unit for batches of about 10,000–15,000 meters (the price being dictated by the modest market in which the company was involved). For purposes of strength (ability to withstand 20-atm inner pressure) and resistance to chemical attack by the fluids flowing through the meter, a special kind of plastic material was chosen. This material is produced by a large company in the United States and (as became evident only during negotiations with the company) can be provided only in the form of the completed article and not as the raw material. As a result costs for production of the flowmeter had to be recalculated, and the decision made under these specific circumstances was to postpone production of the device.

The conclusions we can draw from this specific example can be generalized:

1. When a product or article is redesigned so that less material is required in its manufacture or so that its weight and/or size are reduced, the new product becomes more competitive. The reason is, obviously, the lowering of its selling price as the result of the reduction in material and transportation costs. Sometimes a device must be made more sophisticated before weight and size can be reduced.

2. The potential market for a new article may be satisfactory for a medium-sized company working within a modest market, but completely unprofitable for a large company.

Let us now consider an unusual example. Company X is in the orange business, as are companies Y and Z. What can X do to compete successfully with Y and Z? (Company X enters into the "how" domain.) Make the oranges better? This would, of course, be a step in the right direction, but here the problem of agricultural selection arises: such a path takes years to complete and its outcome is certainly doubtful. Perhaps it would be worthwhile to consider peeled oranges. Perhaps a highly productive machine and process could be designed to "strip" the oranges and wrap them in attractive paper or in plastic. Perhaps the peel could be made edible—sweet or tart. Or perhaps each orange could be packed with a cheap knife for peeling it.

A different approach would entail making the oranges cheaper to the consumer. One way of reducing the price would be to cut down on transportation expenses. For instance, a change from air freight to ships would reduce the expenses considerably, but the problem then arises of how to keep the fruit fresh while being shipped. One Israeli company has found a solution to this type of problem for tomatoes. It has developed a product with an extended shelf life: the tomatoes remain unspoiled for about three months (but this again is a problem of agricultural selection). Another way to reduce the cost of transportation would be better utilization of the cargo's volume: cubic oranges would occupy less space (about 50 percent) in the hold of a ship or aircraft; or cube-shaped eggs packed in plastic sheets not only would save volume, but also would provide a solution to the problem of handling the cargo "like eggs."

As has already been discussed, to capture the market a product

must be cheaper than its competitors, have at least the same quality, and serve the same comparative purposes. One of the most effective means of achieving this aim is to make the product disposable wherever possible and acceptable. It then can be mass produced, and the most effective manufacturing and processing techniques can be used. In this way costs can be reduced, as can the selling price per product unit. Three excellent examples of this type of marketing policy are as follows.

1. Single-use injection syringes and infusion containers can be produced from plastics instead of from glass and metal, since they do not have to stand up to the demands of repeated sterilization. These medical items can thus be mass produced by highly productive casting machines.

2. Many soft-drink containers are now in the form of light metal or plastic cans (instead of glass bottles). Here again, highly productive machines can be used for their manufacture. The low weight of these containers reduces transportation expenses, the fact that they are disposable reduces strength requirements, and washing problems are obviated. Costs can thus be reduced dramatically.

3. The ballpoint pen with a disposable refill is another good example of the manufacture of a cost-effective product. The demand for refills is sufficiently large to warrant completely automatic mass production and assembly of the components of the refills. In this way the expenses per refill become so low that the market is flooded with several kinds of refills produced by several companies.

Another powerful weapon that can be used to capture the market is the development of an article that is more universal than those now available and thus can replace a whole range of products. Consider the following example.

Any automatic or semiautomatic machine used in manufacturing is equipped with one or more feeders, that is, devices that automatically provide the processing units of the machine with raw materials and parts. For many years there was a large variety of feeders on the market—drum, bin, revolving plate, swinging, plunger, sweep-fork. In the 1950s, a feeder based on a new con-

cept—vibration—appeared, and soon replaced almost all the existing types of feeders used in automatic production machines. In this way the market was increased considerably, and the chances for commercial success of the product were similarly improved. The vibrating-feeder concept (see Principle IV) is thus a shining example of the effect that can be achieved by introducing a revolutionary idea for solving a "how" type of problem. Vibrating feeders are much more universal, more flexible, and cheaper than their predecessors.

Let us now look at a different way of market penetration. Many products on the consumer market can be improved or changed so as to obtain new properties by integration of an additional part or device. This is often typical of electronic equipment, but is also true for mechanical and combination items. The creation of such adapters is a useful tool for penetrating a market. Obviously in the initial development stage these adapters belong to the "what" kind of problem.

The "speed warning bleeper" serves as an example of such an adapter. The tremendous number of cars on the roads provides a reasonable source of optimism in the prediction of the marketing fate of the device. The air compressor driven by a car's exhaust gases is another example of such an adapter-type product. Perhaps someone can design an adapter to transform a washing machine into a dishwasher.

In this book we have considered the conceptual design process from the very beginning when the question of "what" to design is first formulated, and have gone through the stages of "how" to solve the specific technical problems. We have also shown the interaction of the design process as creative thinking with the marketing issue. At this point it seems appropriate to wish the reader good luck in the creative process. Should this text prompt some useful associations and help to initiate fruitful ideas, the author will feel deep satisfaction with his endeavors.

Appendix

BIBLIOGRAPHY

1. Altshuller, H.S., *The Algorithm of an Invention* (in Russian), Moskovskij Rabochy, Moscow, 1973.
2. Hansen, R.F., *Konstruktionssystematik. Grundlagen für Eine Allgemeine Konstruktionslehre,* VEB Verlag Technik, Berlin.
3. Jewkes, J., Sawers, D., and Stillerman, R., *The Sources of Invention,* 2d ed., Norton, New York.
4. Krick, E.V., *An Introduction to Engineering and Engineering Design,* Wiley, New York, 1965.
5. Mayall, W.H. *Principles in Design,* 1st ed., Van Nostrand Reinhold, 1979.
6. Orlov, P.I., *The Basics of Design* (in Russian), Mashynostrojenie, Moscow, 1968.
7. Phal, G., and Beitz, B., *Konstruktonslehre,* Springer Verlag, Berlin, 1976.
8. Rodemacker, Wolf G., *Methodisches Konstruiren,* Springer Verlag, Berlin, 1976.
9. Hubka, V., *Theorie der Konstruktionsprocesse,* Springer Verlag, Berlin, 1976.
10. Artobolevsky, I.I., *Mechanisms in Modern Engineering Design,* 6 vols. (translated from Russian by N. Weinstein), MIR Publishers, Moscow, 1975.
11. Bellman, R. *An Introduction to Artificial Intelligence: Can a Computer Think?* Boyd & Fraser, San Francisco, 1978.
12. Rayen, D. L. *Computer-aided Kinetics for Machine Design.* Marcel Dekker, New York, 1981.

References 1–5 and 7–9 constitute excellent reading material in the field of conceptual and creative design and thinking. Reference 6 is an excellent source of the knowledge, rules, and ideas of design and of solution choices in the "how" direction (unfortunately it is available only in Russian). Reference 10 is a superlative collection of kinematic, mechanical, hydraulic, pneumatic, and electromagnetic solutions. References 11 and 12 discuss the role of the computer.

RECOMMENDED READING

The following list gives additional material on different aspects of creative thinking and conceptual design.

Archer, B., "A View of the Nature of Design Research," in *Design: Science: Method* (R. Jacques and J.A. Powell, eds.)., Westbury House, 1981.

Asimow, M., *Introduction to Design,* Prentice-Hall, Englewood Cliffs, N.J., 1966.

Bowman, W. *Graphic Communication,* Wiley, New York, 1968.

Burstall, A.F., *A History of Mechanical Engineering,* Faber & Faber, 1963, 86 pp.

Chestnut, H. *System Engineering Methods,* Wiley, New York, 1967.

Cooper, R.G., "Project New Product, Factors in New Product Success," *Europ. J. Marketing,* 14:277–292, 1980.

De Bono, E., *Teaching Thinking,* Temple Smith, London, 1976.

Faradane, J., *Information for Design. Design Method,* Buttersworth, London, 1966.

French, M.J., *Engineering Design, the Conceptual Stage,* Heinemann, London, 1971.

Glegg, G.L., *The Science of Design,* Cambridge University Press, Cambridge, England, 1973.

Gordon, W.J.J., *Synectics. The Development of Creative Capacity,* Harper & Row, New York, 1961.

Gregory, S.A., *The Design Method,* Buttersworth, London, 1966.

Gregory, S.A., "What We Know About Designing and How We Know It. Current Design Thinking", *I. Chem. E. Midlands,* 815:1–13, 1979.

Hubka, V., *Principles of Engineering Design,* Buttersworth, London, 1982.

Jones, J.C., *Design Methods, Seeds of Human Futures,* Wiley, New York, 1970.

Jones, J.C., *Design Methods,* Wiley, New York, 1980.

Jones, J.C. and Thornley, D.G., *Conference on Design Methods,* Pergamon, Oxford, 1963.

Kardos, G., in *Planning and Creating Successful Engineering Designs* (S.F. Love, ed.), Van Nostrand, New York, 1980.

Krick, E.V., *An Introduction to Engineering Methods, Concepts and Issues,* Wiley, New York, 1976.

Lenat, D.B., "An Artificial Intelligence Approach to Discovery in Mathematics as Heuristic Search," in *Knowledge-Based Systems in AI* (R. Davis and D. Lenat, eds.), McGraw-Hill, New York, 1982.

Love, F., *Planning and Creating Successful Engineering Design,* Van Nostrand Reinhold, New York, 1980.

Marples, D.L., *The Decisions of Engineering Design,* Institute of Engineering Designers, London, 1960.

Morrison, D., *Engineering Design, the Choice of Favourable Systems,* McGraw-Hill, London, 1968.

Nilson, N.J., *Principles of Artificial Intelligence,* Tioga, Palo Alto, Calif., 1980.

Osborn, A.F., *Applied Imagination,* Scribner's, New York, 1963.

Ostrofsky, B. *Design, Planning and Development Methodology,* Prentice-Hall, Englewood Cliffs, N.J., 1977.

Pugh, S., "Concept Selection—The Ability to Compete," presented at Design Policy Conference at the Royal College of Art, London, July 1982.

Pugh, S., "Concept Selection—A Method That Works," *Proceedings International Conference Engineering Design* (ICED 81), Rome, 1981, pp. 497–506.

Pugh, S., and Smith, D.G., "The Danger of Design Methodology," presented at First European Design Research Conference, Changing Design, Portsmouth Polytechnic, 1976.

Roe, P.H., Soulis, G.N., and Handa, V.K., *The Discipline of Design*, Allyn & Bacon, Boston, 1967.

Roth, K., *Konstruiren mit Konstruktion-katalogen*, Springer Verlag, Berlin, 1982.

Rzevski, G., "On the Design of a Design Methodology," in *Design: Science: Method* (R. Jacques and J.A. Powell, Eds.), Westbury House, 1981.

Vesper, K.H., *Engineers at Work: A Case Book*. Houghton Mifflin, Boston, 1975.

Wallace, K.M., "Engineering Design Research," in *Design: Science: Method* (R. Jacques and J.A. Powell, eds.), Westbury House, 1981.

Wallace, P.J., *The Techniques of Design*, Pitman, London, 1952.

Wells, G.L., "A Study of Some Process Heuristics—Current Design Thinking," *In Chem. E. Midlands*, 612:1-10, 1979.

Weinberg, M. *An Introduction to General Systems Thinking*, Wiley, New York, 1975.

INDEX